9 Springer Series in Solid-State Sciences

Edited by Hans-Joachim Queisser

Springer Series in Solid-State Sciences

Editors: M. Cardona P. Fulde H.-J. Queisser

Volume 1 **Principles of Magnetic Resonance** 2nd Edition
By C. P. Slichter

Volume 2 **Introduction to Solid-State Theory**
By O. Madelung

Volume 3 **Dynamical Scattering of X-Rays in Crystals**
By Z. G. Pinsker

Volume 4 **Inelastic Electron Tunneling Spectroscopy**
Editor: T. Wolfram

Volume 5 **Fundamentals of Crystal Growth I.** Macroscopic Equilibrium and Transport Concepts
By F. Rosenberger

Volume 6 **Magnetic Flux Structures in Superconductors**
By R. P. Huebener

Volume 7 **Green's Functions in Quantum Physics**
By E. N. Economou

Volume 8 **Solitons and Condensed Matter Physics**
Editors: A. R. Bishop and T. Schneider

Volume 9 **Photoferroelectrics**
By V. M. Fridkin

Volume 10 **Phonon Dispersion Relations in Insulators**
By H. Bilz and W. Kress

Volume 11 **Electron Transport in Compound Semiconductors**
By B. R. Nag

Volume 12 **The Physics of Elementary Exitations**
By S. Nakajima

Vladimir M. Fridkin

Photoferroelectrics

With 63 Figures

Springer-Verlag Berlin Heidelberg New York 1979

Professor Dr. Vladimir M. Fridkin

Institute of Crystallography, Academy of Sciences of the USSR,
Leninsky prospect 59, Moscow 117333, USSR

Series Editors:

Professor Dr. Manuel Cardona
Professor Dr. Peter Fulde
Professor Dr. Hans-Joachim Queisser

Max-Planck-Institut für Festkörperforschung
Heisenbergstrasse 1, D-7000 Stuttgart 80, Fed. Rep. of Germany

ISBN-13:978-3-642-81353-5 e-ISBN-13:978-3-642-81351-1
DOI: 10.1007/978-3-642-81351-1

Library of Congress Cataloging in Publication Data. Fridkin, Vladimir Mikhaĭlovich. Photoferroelectrics. (Springer series in solid-state sciences ; v. 9) Includes bibliographical references and index. 1. Photoferroelectric effect. 2. Ferroelectric crystals. I. Title. II. Series. QC715.17.F74 537.2'44 79-13366

2153/3130-543210

To Nadia

Preface

The investigation of nonequilibrium electronic processes in ferroelectric crystals has proved to be of some importance from both the theoretical and technological standpoints over the last ten years. The study of photoferroelectrics has made important contributions to the physics of ferroelectrics in general, and has also provided a new range of materials for optical storage by holography. This monograph is an attempt to summarize the main results with emphasis on the physics of these phenomena. Particular attention is given to the work of Soviet physicists, with whose work I happen to have been more closely associated.

The author wishes to express his deep appreciation to a number of persons who cooperated with him in the work on this book. These are, in the first place, a group of researchers and post-graduates at the Institute of Crystallography of the USSR Academy of Sciences, with whom the author has been working for a number of years. These investigations have been conducted in close contact with A.A. Grekov and A.I. Rodin (Rostovon-Don University). Discussions with A. Glass (Bell Laboratory, Murray Hill), G. Chanussot and G. Godefroy (Dijon University), F. Micheron (Thomson, Orsay), W. Ruppel (University of Karlsruhe), and D. von der Linde (Max-Planck-Institut, Stuttgart) were of great help. The author wishes to extend his cordial thanks to his close associates K.A. Verkhovskaya and L.I. Savina, whose aid in the work on the manuscript was invaluable.

Moscow, May 1979 V.M. Fridkin

Contents

1. Introduction

Since Valasek's discovery of the ferroelectric properties of Rochelle salt nearly 60 years ago, ferroelectricity has been regarded as one of the traditional branches of dielectric physics. It has had important applications in lattice dynamics, quantum electronics, and nonlinear optics.

The study of electron processes in ferroelectrics was begun with VUL's investigations of the ferroelectric properties of barium titanate [1.1]. Intrinsic and extrinsic optical absorption, band structure, conductivity and photoconductivity, carrier mobility, and transport mechanisms were examined in this compound, and in other perovskite ferroelectric semiconductors. An important discovery was that of the highly photosensitive photoconducting ferroelectrics of type $A^{V}B^{VI}C^{VIII}$ (e.g. SbSI) by MERZ et al. in 1962 [1.2,3].

A large number of ferroelectric semiconductors (some photosensitive, some not) are now known, including broad-band materials (e.g. lithium niobate, lithium tantalate, barium and strontium niobate, and type-$A_2^{V}B_3^{VI}$ compounds), and narrow-band semiconductors (e.g. type-$A^{IV}B^{BI}$ compounds). A series of improper ferroelectric semiconductors and photosensitive ferroelastics have been discovered, of which Sb_5O_7I is an example.

Owing to the uncertainty of their band structure, the difficulty in determining the nature of the levels, the complexity of alloying, and their generally low mobility values, ferroelectrics are rarely of interest regarded as nonlinear semiconductors. The most fruitful approach has been the study of the influence of electrons (especially nonequilibrium electrons) and electron excitations on phase transitions and ferroelectric properties. A large group of phenomena have recently been discovered and investigated.

One important photoferroelectric phenomenon is the effect of nonequilibrium conductivity on the birefringence of ferroelectrics and electro-optical crystals – the photorefractive effect. This has been applied to three-dimensional phase holography, where some of these materials compete with silver halides in sensitivity and resolution. Investigation of the photorefractive effect led to the discovery of a new mechanism of charge transfer in ferroelectrics, the anomalous photovoltaic effect. A homogeneous photoferroelectric crystal might well be a possible solar energy converter.

This book is an attempt to survey photoferroelectric phenomena, and deliberately avoids incursions into related branches of ferroelectric physics that have been covered by a number of comprehensive monographs which have recently appeared [1.4-14]. For example, such closely related subjects as band structure and optical and electrical absorption are omitted: these are covered to some extent in [1.14]. Neither has any attempt been made to give a systematic description of the applications of these phenomena; those of the photorefractive effect alone would be worth a separate monograph.

Another limitation of the present work springs from the present state of incompleteness of the subject and the fact that this survey was undertaken during a period of intensive development. The author did not even attempt to present the subject as a whole, and thus could not avoid some one-sidedness. This, in the author's opinion, is both the merit and demerit of the book.

The starting-point of this discussion is the thermodynamics of photoferroelectrics, considered in Chapter 2. The effect of nonequilibrium electrons on the properties of ferroelectrics is discussed on the basis of the phenomenological theory of ferroelectric phase transitions. In Chapter 3, the microscopic approach to the ferroelectric phenomena is developed. Chapter 4 deals with the screening of the spontaneous polarization by the nonequilibrium carriers. Experimental results on photoferroelectric phenomena are discussed in Chapter 5, including the photostimulated shift of the Curie point, the photodomain effect, and the influence of nonequilibrium electrons on the spontaneous polarization. Chapter 6 describes the results of the investigation of the anomalous photovoltaic effect in ferroelectrics, which can be used to advantage in creating new energy conversion elements. The next chapter is devoted to the photorefractive effect in ferroelectrics and its applications to three-dimensional hologram recording. In conclusion, Chapter 8 summarizes the experimental data on the influence of nonequilibrium carriers on screening phenomena in ferroelectrics.

2. The Thermodynamics of Photoferroelectrics

Phase transitions in solids are now believed to be associated with instability of one of the vibration modes. In particular, ferroelectric phase transitions are associated with the unstable ('soft') optical mode at the center of the Brillouin zone. While the behavior of the electron spectrum and electron-phonon interactions have practically been ignored until recently, a large body of experimental data has been accumulated on the effect of electrons, and of nonequilibrium electrons in particular, on ferroelectric phase transitions and the properties of ferroelectrics well below the Curie point. The effect of nonequilibrium electrons on phase transitions and properties of ferroelectrics is of great interest from the experimental standpoint. Phase transitions due to such electrons are described as photostimulated in the literature, and the phenomena associated with the effect of nonequilibrium electrons (and of electron excitations generally) on the ferroelectric properties, as photoferroelectric [2.1-4]. Thus, we shall use the term photoferroelectrics to refer to crystals which change their ferroelectric properties (or properties connected with the spontaneous polarization in one way or another) when optically excited in the intrinsic or extrinsic region of the spectrum.

While photostimulated phase transitions can occur in all semiconductors that undergo phase transitions, investigation of photoferroelectrics has the advantage that it permits photoferroelectric phenomena to be correlated with the basic phenomenological parameters of ferroelectrics, such as the Curie-Weiss constant, the spontaneous polarization, the heat-capacity discontinuity, and other independently measured parameters.

The existence of photostimulated phase transitions in solids, particularly in ferroelectrics, is of fundamental interest. When considering the mechanism of phase transitions for a crystal, the contribution that the electron subsystem makes to the total free energy of the crystal is usually neglected, and thus it is assumed that the phase transition is not associated with the electron excitation in the crystal. To see why this assumption is made, consider the Gibbs-Helmholtz equation

$$F = E + T\frac{\partial F}{\partial F} \quad , \tag{2.1}$$

where F is the free energy, and E is the internal energy. Integrating, we get

$$F = -T\int_0^T \frac{E}{T^2}\,dT \quad . \tag{2.2}$$

The relation between the internal energy E and the heat capacity C_V is

$$E(T) = E(0) + \int_0^T C_V(T)\,dT \quad . \tag{2.3}$$

Substituting (2.3) into (2.2), we obtain the following expression for the free energy:

$$F = E(0) + T\int_0^T \frac{dT}{T^2}\int_0^T C_V(\tau)\,d\tau \quad . \tag{2.4}$$

Equation (2.4) shows that the free energy of the crystal is completely determined by the heat capacity as a function of temperature. Thus, the importance of the contribution of the electrons to the free energy is determined by the ratio of the electron heat capacity $C_V^{e\ell}$ to the lattice heat capacity C_V^L.

Above the Debye temperature, the general expressions for this ratio are

$$\frac{C_V^{e\ell}}{C_V^L} = \frac{N_c}{n_0} \ll 1 \tag{2.5}$$

for a nongenerate semiconductor or dielectric [where $N_c = (2\pi m^* kT/h^2)^{3/2}$ is the density of states of an electron or a hole in the band, m* is the effective mass of the electron or hole, and n_0 is the number of atoms per cubic centimeters], and

$$\frac{C_V^{e\ell}}{C_V^L} \simeq \frac{kT}{E_F} \ll 1 \tag{2.6}$$

for a metal, where the Fermi energy E_F is of the order of several eV, and $kT \simeq 0.025$ eV at room temperature. Thus, for both metals and semiconductors, the contribution of the electrons to the free energy of the crystal can be neglected above the Debye temperature. Thus, it would appear that the contribution of the electrons can become substantial only near $T = 0$, where C_V tends to zero as T^3.

However, the electron subsystem may contribute substantially to the free energy of the crystal near the phase transition temperature in addition to the region around absolute zero. This comes about because the electrons may make an appreciable contribution, not to the heat capacity itself, but it its anomalous part which arises during the phase transition. This conclusion was first suggested in [25] for ferroelectric phase transitions within the framework of the phenomenological theory of Landau, Ginsburg, and Devonshire, and was developed in [2.6,7]. The effect of electron excitations on phase transitions of a different nature has subsequently been studied. In this chapter we shall analyze photoferroelectric phenomena and the effect of electrons on the properties of ferroelectrics on the basis of the phenomenological theory of ferroelectric phase transitions.

2.1 The Free Energy of a Ferroelectric Semiconductor

The relatively high concentration of carriers in a ferroelectric semiconductor necessitates taking into account the free energy of the electron subsystem in the expression for the free energy of the crystal near the phase transition temperature. We assume firstly that the free energy of the electron subsystem, F_2, is everywhere (except in the vicinity of the Curie point) low compared with the free energy of the lattice, F_1, and that the phase transition itself is associated with instability in the lattice vibrations (otherwise we have a purely electron phase transition, as in the vanadium oxides). Secondly, we shall deal only with proper ferroelectrics for which (according to [2.8]) the free lattice energy F_1 near the Curie point can be expanded into a series in even powers of the spontaneous polarization P. Finally, we shall neglect the anisotropy of the dielectric properties and assume that crystal polarization takes place only in the direction of the spontaneous polarization axis (c axis).

The free energy of the crystal, F, can be represented as a sum of the free energies of the lattice in the paraelectric region, F_0, and in the ferroelectric region, F_1, and the free energy of the electron subsystem, F_2 [2.5-7]:

$$F = F_0 + F_1 + F_2 \quad , \tag{2.7}$$

where

$$F_0(T) = F(P = 0,\ \sigma_k = 0,\ N_i = 0) \quad , \tag{2.8}$$

$$F_1 = \frac{1}{2}\alpha P^2 + \frac{1}{4}\beta P^4 + \frac{1}{6}\gamma P^6 - \frac{1}{2}\sum_i \sum_k s_{ik}\,\sigma_i\sigma_k - P^2 \sum_k \nu_k\sigma_k \quad , \tag{2.9}$$

$$F_2 = \sum_i N_i\, E_i(T,\ P,\ \sigma_k) \quad . \tag{2.10}$$

Here α, β, and γ are known coefficients in the expansion of the free energy in the polarization P [1.13,2.9-10]; σ_k are the components of the mechanical stress tensor; $s_{ik} = s^P_{ik} = \partial^2 F_1/\partial\sigma_i\partial\sigma_k$ are the components of the elastic constant tensor; and $\nu_k = \partial^3 F_1/\partial P^2\partial\sigma_k$ are the components of the electrostriction tensor. The summation in (2.10) is over energy levels of the electrons, labelled by i. In the expression (2.10) for the free energy of the election subsystem E_i and N_i are the energies of the levels and the respective electron (hole) concentrations in the crystal. The configuration portion of the free energy has been neglected in (2.10) for the time being, and in (2.9) the polarization has been taken to be uniform and the correlation term ignored. In the general case, (2.9) must be supplemented by the Poisson equation (see Chapter 4). The expression (2.10) describes the contribution to the free energy exclusively from free electrons and holes, for which the effect of the electron states on the spatial distribution of the polarization P can be neglected, so, strictly speaking, it cannot be applied to the case of deep local electron levels (see Chap.3).

To be specific, let us consider an n-type ferroelectric semiconductor, whose energy-band diagram is presented in Fig. 2.1. Assume that a crystal with an energy gap E_g and a free-electron concentration n has electron traps of one type (with energy u_1, concentration M, and electron concentration at these levels N) and hole traps of one type (with energy u_2, and hole concentration at these levels p). Neglecting the contribution of the recombination

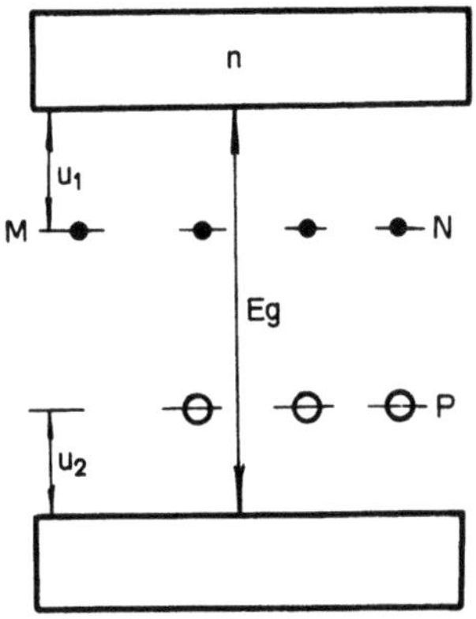

Fig. 2.1. Energy-band diagram of a ferroelectric semiconductor

levels and assuming that N, p >> n, as is usually the case in high-resistance semiconductors, we can represent the free energy F_2 as

$$F_2 = nE_g + N(E_g - u_1) - pu_2 \simeq N(E_g - u_1 - u_2) \equiv N\tilde{E} \quad , \tag{2.11}$$

where $\tilde{E}$ stands for $E_g - u_1 - u_2$, and the condition of electrical neutrality, $p = n + N$, has been used. Assuming further that the function F_2 must be invariant under the same symmetry transormation as the function F_1, the energy $\tilde{E}$ can be expanded near the phase transition temperature in a series in P and δ_k as follows:

$$\begin{aligned} \tilde{E}(T, P, \sigma_k) = \tilde{E}_0(T) &+ \frac{1}{2} (\tilde{E}_P^{II})_0 P^2 + \frac{1}{4} (\tilde{E}_P^{IV})_0 P^4 + \frac{1}{6} (\tilde{E}_P^{VI})_0 P^6 \\ &+ \sum_k \tilde{E}'_k \sigma_k + \frac{1}{2} \sum_k \sum_i \tilde{E}''_{ki} \sigma_k \sigma_i + P^2 \sum_k \tilde{E}'''_k \sigma_k \quad , \end{aligned} \tag{2.12}$$

where we have adopted the following notation:

$$\begin{aligned} &(\tilde{E}_P^{II})_0 = \left(\frac{\partial^2 \tilde{E}}{\partial P^2}\right)_0 = a \quad , \quad (\tilde{E}_P^{IV})_0 = \left(\frac{\partial^4 \tilde{E}}{\partial P^4}\right)_0 = b \quad , \quad (\tilde{E}_P^{VI})_0 = \left(\frac{\partial^6 \tilde{E}}{\partial P^6}\right)_0 = c \quad , \\ &\tilde{E}'_k = \left(\frac{\partial \tilde{E}}{\partial \sigma_k}\right)_0 \quad , \quad \tilde{E}''_{ki} = \left(\frac{\partial^2 \tilde{E}}{\partial \sigma_k \partial \sigma_i}\right)_0 \quad , \quad \tilde{E}'''_k = \left(\frac{\partial^3 \tilde{E}}{\partial P^2 \partial \sigma_k}\right)_0 \quad . \end{aligned} \tag{2.13}$$

Combining (2.7-13), we obtain the following expression for the free energy of a ferroelectric semiconductor:

$$F(T,\, P,\, \sigma_k,\, N) = F_{ON} + \frac{1}{2}\,\alpha_N\, P^2 + \frac{1}{4}\,\beta_N\, P^4 + \frac{1}{6}\,\gamma_N\, P^6$$
$$+ N \sum_k \tilde{E}_k'\, \sigma_k - \frac{1}{2} \sum_i \sum_k s_{Nik}\, \sigma_i \sigma_k - P^2 \sum_k \nu_{Nk} \sigma_k \quad , \tag{2.14}$$

where the coefficients are defined as

$$F_{ON} = F_0 + N\tilde{E}_0 \quad , \quad \alpha_N = \alpha + aN \quad , \quad \beta_N = \beta + bN \quad ,$$
$$\gamma_N = \gamma + cN \quad , \quad \nu_{Nk} = \nu_k - \tilde{E}_k'''N \quad , \quad s_{Nik} = s_{ik} - \tilde{E}_{ik}''N \quad . \tag{2.15}$$

Equation (2.14) must be supplemented by two equations of state

$$\frac{\partial F}{\partial P} = \mathscr{E} = \alpha_N P + \beta_N P^3 + \gamma_N P^5 - 2P \sum_k \nu_{Nk} \sigma_k \quad , \tag{2.16}$$

$$\frac{\partial F}{\partial \sigma_k} = -u_k = N\tilde{E}_k' - \frac{1}{2} \sum_i s_{Nik}\, \sigma_i - P^2 \nu_{Nk} \quad , \tag{2.17}$$

where $\mathscr{E}$ is the electric field, and u_k are components of the deformation tensor.

2.2 First- and Second-Order Phase Transitions

Before proceeding to analyze the system of equations (2.14,16,17) and the effect of the electrons on the phase transition and the ferroelectric properties of the crystal, we review the main features of the thermodynamic theory of ferroelectrics [1.13,2.8-2.11]. For simplicity we shall assume that $N = \mathscr{E} = 0$, $\sigma_k = 0$. We consider separately the cases of phase transitions of first and second order.

For a second-order phase transition it can be shown [1.13,2.9] that terms of order P^6 or higher in the expression for the free energy (2.14) can be neglected, so we consider the expression

$$F = F_0 + \frac{1}{2}\alpha P^2 + \frac{1}{4}\beta P^4 \quad . \tag{2.18}$$

The value of P in termal equilibrium is obtained by requiring the free energy to have a minimum; i.e.,

$$\frac{\partial F}{\partial P} = \alpha P + \beta P^3 = 0 \quad , \tag{2.19}$$

$$\frac{\partial^2 F}{\partial F^2} > 0 \quad . \tag{2.20}$$

The two solutions of equation (2.19), $P = 0$, and $P^2 = -\alpha/\beta$, correspond to the paraelectric phase and to the ferroelectric phase, respectively. From the condition (2.20)

$$\alpha + 3\beta P^2 > 0 \tag{2.21}$$

it follows that in the paraelectric region ($P = 0$) α must be positive, and in the ferroelectric region ($P^2 = -\alpha/\beta$) α must be negative. Denoting the phase transition temperature, or Curie point, by T_0, expanding α in a series in $(T - T_0)$ near $T = T_0$ and restricting ourselves to the linear term, we can represent α in the form

$$\alpha = \alpha_T^{\prime}(T - T_0) \equiv \frac{2\pi}{C}(T - T_0) \tag{2.22}$$

where C stands for the Curie-Weiss constant. According to (2.21) β is positive everywhere below the Curie point. We assume that β is independent of temperature. Thus, the temperature dependence of the spontaneous polarization below the Curie point is given by

$$P^2 = \frac{\alpha_T^{\prime}(T_0 - T)}{\beta} \quad , \; T < T_0 \quad . \tag{2.23}$$

From the equation for the entropy,

$$S = -\frac{\partial F}{\partial T} = S_0 + \frac{(\alpha_T')^2}{\beta}(T - T_0) \quad , \tag{2.24}$$

where $S_0 = -\partial F_0/\partial T$ is the entropy of the paraelectric phase, we can determine the heat-capacity jump $C_p = T(\partial S/\partial T)_p$ in a second-order phase transition; we get

$$\Delta C_p = \frac{T_0}{\beta}(\alpha_T')^2 \quad . \tag{2.25}$$

In conformity with the general nature of second-order phase transitions, a finite heat-capacity jump occurs at the Curie point, and the latent heat is zero.

To determine the temperature dependence of the dielectric constant ε near the Curie point, we take $\mathscr{E} \neq 0$ in (2.16). We then obtain [2.9] in the paraelectric region ($T > T_0$),

$$\varepsilon \simeq \frac{2\pi}{\alpha} \quad , \tag{2.26}$$

and in the ferroelectric region ($T < T_0$)

$$\varepsilon \simeq -\frac{\pi}{\alpha} \quad . \tag{2.27}$$

The expression (2.26) can be combined with (2.22) to give the temperature dependence of ε in the paraelectric phase in the form known as the Curie-Weiss law,

$$\varepsilon \simeq \frac{C}{T-T_0} \tag{2.28}$$

Figure 2.2 shows the dependence of the free energy on the polarization, $F = F(P)$, the temperature dependence of the square of the spontaneous polarization, and the temperature dependence of the reciprocal dielectric constant near a second-order ferroelectric phase transition. The ratio of the slopes of the lines of ε^{-1} against T for the ferroelectric and paraelectric phases is equal to minus two, as follows from (2.26,2.27).

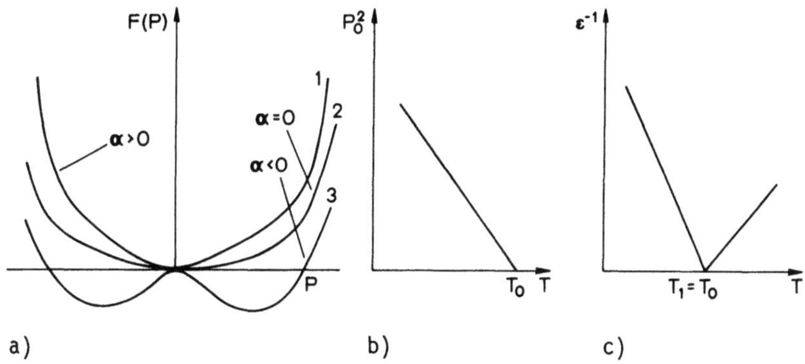

Fig. 2.2. Dependences F(P)(a), $P_0^2(T)$(b), and $\varepsilon^{-1}(T)$(c) near a second-order phase transition

A first-order ferroelectric transition is described by the expansion (2.14), where the expansion coefficients do not all have the same sign in the paraelectric region. In this case it is necessary to take into account all the three terms of the expansion in P. We take it that in the expansion (2.14) $\beta < 0$, $\gamma > 0$, and α satisfies (2.22). As follows from (2.16) for $\mathscr{E} = 0$, in this case the free energy F has two different minima for the temperature $T = T_1$ corresponding to two different values of P. One of these minima corresponds to P = 0, and the other, to $P = P_0 \neq 0$. Thus, the temperature $T = T_1$ is the temperature of a first-order phase transition at which the crystal changes discontinuously from one stable state to another. The temperature dependence P = P(T) in the ferroelectric phase is determined from the condition that free energy has a minimum (2.16)

$$\alpha P + \beta P^3 + \gamma P^5 = 0 \quad , \tag{2.29}$$

whence

$$P^2 = -\frac{\beta}{2\gamma}\left[1 + \left(1 - \frac{4\alpha\gamma}{\beta^2}\right)^{1/2}\right] \quad . \tag{2.30}$$

The condition for the equality of the free energies F at P = 0 and $P = P_0$ is

$$\frac{1}{2}\alpha P^2 + \frac{1}{4}\beta P^4 + \frac{1}{6}\gamma P^6 = 0 \tag{2.31}$$

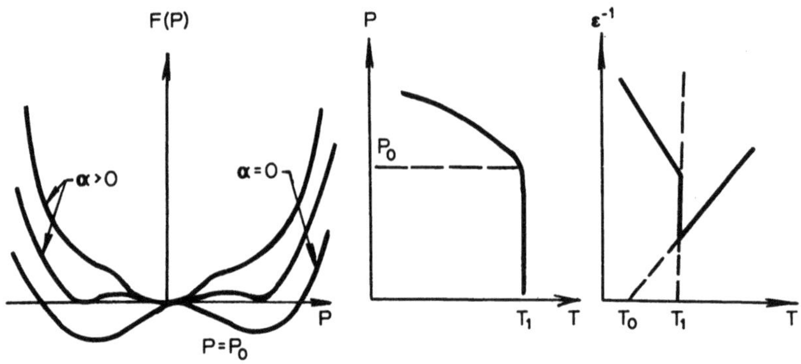

a) b) c)

Fig. 2.3. Dependences F(P)(a), P(T)(b), and ε^{-1}(T)(c) near a first-order phase transition

The values of P_0 and α at the phase transition $T = T_1$ are obtained by requiring the conditions (2.29) and (2.31) to hold simultaneously. We obtain

$$P_0^2 = P^2(T_1) = \frac{3}{4}\left(-\frac{\beta}{\gamma}\right) \quad , \tag{2.32}$$

$$\alpha(T_1) = \frac{3}{16}\frac{\beta^2}{\gamma} \quad . \tag{2.33}$$

Figure 2.3 illustrates, for a first-order phase transition, the dependence of the free energy on the polarization, $F = F(P)$, the temperature dependence of the spontaneous polarization $P = P(T)$, and the temperature dependence of the reciprocal dielectric constant.

Calculating the entropy using (2.24) shows that a first-order phase transition is accompanied by the release of latent heat, and an entropy discontinuity of

$$\Delta S = \frac{2\pi}{C} P_0^2 \tag{2.34}$$

Another feature of a first-order phase transition is the temperature hysteresis. From Fig. 2.3a, it can be seen that the transition of a crystal from the equilibrium position at $P = 0$ to the equilibrium position at $P = P_0$ involves the forcing of an energy barrier, whose height is proportional to $|\beta|^3/\gamma^2$. This barrier causes the temperature hysteresis of the phase transition. The value of the temperature hysteresis, ΔT_h, can be estimated as the difference of the phase transition temperature T_1 and the Curie-Weiss temperature T_0 and determined from (2.33) and (2.22) as

$$\Delta T_h \simeq T_1 - T_0 = \frac{1}{8\pi} C \frac{\beta^2}{\gamma} \quad . \tag{2.35}$$

2.3 Effect of Electrons on Phase Transitions: Photoferroelectric Phenomena

Let us now consider the system (2.14-17) for $N \neq 0$. The free energy of the electron subsystem is taken account of by renormalizing the coefficients of expansion of the free energy in the parameter P according to (2.15), which results in changes in the thermodynamic functions, and may alter the nature of the phase transition. We consider some of these effects.

a) The Curie-Point Shift. It is clear from the relation (2.15) for α_N that the Curie point, defined by (2.22), will be shifted by the electrons. Denoting by T_{0N} the Curie point in the presence of electrons with a concentration N, and by T_0 the Curie point in the absence of electrons, we have

$$T_{0N} - T_0 = \Delta T_N = - \frac{C}{2\pi} aN \quad , \tag{2.36}$$

where C is the Curie-Weiss constant and $a \equiv (\partial^2 \tilde{E}/\partial P^2)$ is the coefficient in the expansion of the energy $\tilde{E}$ in polarization (2.12). From the condition that the free energy have a minimum, $(\partial^2 F/\partial P^2)_0 = Na > 0$, it follows that $a > 0$, and hence the electrons lower the Curie point; the shift being proportional to the electron concentration N. As noted in Sect. 2.1, in high-resistance ferroelectric semiconductors, N stands for the electron (hole) concentration in traps. When all the traps are filled (N = M), the dependence of the Curie point shift ΔT_N on N must show saturation. Then (2.36) holds only for values of ΔT_N small enough for the Landau-Ginsburg-Devonshire expansion to be applicable.

b) Change in Spontaneous Polarization. The change in spontaneous polarization due to the electrons can be obtained from (2.30) and (2.15)

$$P_N^2 = P^2(T)\left[1 + \frac{bNP^2(T)}{\beta P^2(T)+2\alpha} + \frac{cN\alpha P^4(T)}{\beta P^2(T)+2\alpha} - \frac{CN}{\gamma}\right] \quad . \tag{2.37}$$

The change in the jump $P_0 = P_0(T_1)$ at the point of the first-order phase transition follows from (2.32) and (2.15)

$$P_{0N}^2 = P_0^2\left[1 + \frac{bN}{\beta} - \frac{cN}{\gamma}\right] \quad . \tag{2.38}$$

If we assume that $b > 0$ and $c > 0$, then, as follows from (2.37,38), the electrons reduce the spontaneous polarization over the entire temperature range of the ferroelectric phase, including the phase transition point itself.

Due to the recharging of deep centers in a ferroelectric, the sign and magnitude of ΔP_0 may depend on the parameters of these centers (see Chap. 3).

c) Photorefraction. The effect of nonequilibrium electrons on the spontaneous polarization described by (2.37,38) results in a change of the spontaneous birefringence $n_i = n_e - n_0$. This is known as the photorefractive effect, or 'optical damage'.

d) Change in Temperature Hysteresis. First-order phase transitions exhibit temperature hysteresis, the microscopic mechanism for which is associated with a nucleation process [2.12]. The value of the temperature hysteresis is proportional to the energy of formation of the critical nucleus. We have seen that this energy is proportional to $|\beta|^3/\gamma$ for ferroelectric phase transitions. If we assume that $b > 0$ and $c > 0$, the electrons reduce the energy $|\beta|^3/\gamma$ and, hence, the change in temperature hysteresis due to the electrons, estimated with the aid of (2.15,35) is

$$\Delta T_{hN} \simeq \frac{3}{32\pi} C \frac{\beta^2}{\gamma}\left[1 + \frac{2bN}{\beta} - \frac{cN}{\gamma}\right] \quad . \tag{2.39}$$

Comparing (2.35), we see that $\Delta T_{hN} < \Delta T_h$ when $b > 0$ and $c > 0$.

e) Change in Spontaneous Deformation. The effect of the electrons on the spontaneous deformation follows from the equation of state (2.17). Putting $\sigma_i = 0$, we have

$$u_{kN} = \nu_{kN} P_0^2 - N\tilde{E}_k' = u_k \left[1 + \frac{\Delta P_{0N}^2}{P_0^2} - \frac{\tilde{E}_k''' N}{\nu_k}\right] - N\tilde{E}_k' \quad , \tag{2.40}$$

where

$$u_k = \nu_k P_0^2 \quad , \quad \Delta P_{0N}^2 = (P_{0N} - P_0)^2 \quad .$$

The effect of the electrons on the volume of a unit cell during phase transition can be estimated similarly. The change is given by

$$\frac{\Delta \upsilon_N}{\upsilon} = \frac{\Delta \upsilon}{\upsilon}\left[1 + \frac{\Delta P_{0N}^2}{P_0^2} - N \sum_k \frac{\tilde{E}_k'''}{\nu_k} - N \sum_k \frac{\tilde{E}_k'}{\nu_k P_0^2}\right] \quad . \tag{2.41}$$

f) Change in Dielectric and Piezoelectric Properties. Combining (2.26,33, 35) leads to the following equation relating the dielectric constant at the point of the first-order phase transition with the temperature hysteresis:

$$\varepsilon(T_1)\Delta T_h \simeq C \quad . \tag{2.42}$$

If we assume that the electrons do not affect the Curie-Weiss constant C and that they reduce the temperature hysteresis, then $\varepsilon(T_1)$ increases, i.e. $\varepsilon_N(T_1) > \varepsilon(T_1)$.

The effect of electrons on the piezoelectric moduli is described by the equation of state (2.17) for $\mathscr{E} \neq 0$:

$$\left(\frac{\partial u_k}{\partial \mathscr{E}}\right)_N = d_{kN} = \frac{\varepsilon_N}{2\pi} \nu_{Nk} P_N \quad . \tag{2.43}$$

g) Change in Latent Heat and in Heat-Capacity Jump. From (2.25,34) we obtain the dependence of the entropy jump ΔS_N and the heat capacity jump ΔC_N on the electron concentration N:

$$\Delta S_N = \frac{2\pi}{C} P_0^2 \left[1 + \frac{bN}{1+\beta} - \frac{cN}{\gamma}\right] \quad , \tag{2.44}$$

$$\Delta C_N = \frac{4\pi^2 T_0}{C^2 \beta}\left[1 - \frac{bN}{\beta}\right] \quad . \tag{2.45}$$

From the equations of state (2.16,17) we can also obtain other equations relating the characteristics of phase transitions with the electron concentration. One common feature of these results is that the increase in electron concentration under the above assumptions ($b > 0$, $c > 0$) results in $|\beta_N| < |\beta|$ and $\gamma_N > \gamma$. This changes the nature of the phase transition, shifting it closer to the critical Curie point [2.8,9], and results in a decrease in temperature hysteresis, in the latent heat of transition, etc.

When a ferroelectric is illuminated in the spectral region of its intrinsic or extrinsic photosensitivity there is an increase in the electron (hole) concentration in the band n due to nonequilibrium carriers. This, in turn, increases the carrier concentration in the traps, N. The relationship between n and N is determined, in each particular case, by the trap parameters, trap energies, and capture cross section. Cutting off the 'intrinsic' light results in depletion of the traps because the electrons (holes) are excited to the band and recombine. Thus, when a ferroelectric is illuminated, one will observe a shift of the Curie point and a change in the physical properties of the ferroelectric due to the change in the concentration of the nonequilibrium carriers or, in the general case, to electron excitations. Early on, we described these phenomena as photoferroelectric. As indicated above, the thermodynamics of ferroelectric semiconductors describes photoferroelectric phenomena, although it does not reveal their microscopic mechanism. The latter will be discussed in Chap. 3.

2.4 The Energy-Gap Anomaly

The effect of phase transitions on the energy gap can be analyzed by considering the expression (2.12). When N means the concentration of free electrons, we have $E = E_g$. For simplicity, we shall make this assumption in the general case, although the identification is sometimes inadmissible.

If the crystal is not subject to mechanical stresses, the relation (2.12) thus reduces to

$$E_g = E_{g0} + \frac{1}{2}aP^2 + \frac{1}{4}bP^4 + \frac{1}{6}cP^6 \quad , \qquad (2.46)$$

where E_{g0} is the energy gap in the paraelectric phase. Thus the discontinuity in the first-order transition from the ferroelectric to the paraelectric phase is approximately

$$\Delta E_g \simeq \frac{a}{2}P_0^2 \simeq \frac{a}{4\pi}C\Delta S \quad . \qquad (2.47)$$

In a second-order phase transition, $\Delta E_g = 0$. However, there is a discontinuity in the derivatives of the energy gap $(\partial E_g/\partial T)_p$ and $(\partial E_g/\partial p)_T$.

From (2.46), since $a > 0$, we see that, if an external electric field is applied in the direction of the spontaneous polarization, the energy gap of the ferroelectric is increased, and the edge of the intrinsic optical absorption band is shifted towards the short wavelength end of the spectrum [1.14]. This effect opposes the Franz-Keldysh effect, and is one order of magnitude larger, as was confirmed by early observations on the ferroelectric crystal SbSI [2.13,14].

In writing down the expressions (2.10,11) for the free energy of the electron subsystem, we took the energy gap to be the free energy per electron-hole pair. This definition has been rigorously substantiated in [2.15,16]. KEYES [2.17] developed these thermodynamic ideas using the model of SLATER [2.18], and derived the general relation

$$\left(\frac{\partial E_g}{\partial T}\right)_p \sim C_V \quad . \qquad (2.48)$$

The energy-gap anomaly in ferroelectric phase transitions is a consequence of this. It also follows that there is such a discontinuity in the energy gap or its derivatives in nonferroelectric transitions.

2.5 Electrical Conductivity of Ferroelectric Semiconductors Near the Curie Point

In the expressions for the free energy of the electron subsystem F_2 (2.10, 11) we did not take into account the problem of arranging N electrons with spins of two opposite directions among M traps. Consider an n-type ferroelectric semiconductor with a concentration of donor levels N, at an activation energy E below the band. Since the electrons obey Fermi-Dirac statistics, the expression (2.11) for the free energy of the electron subsystem should be modified as follows [2.19]:

$$F_2 = nE - nkT\left[\ln\left(\frac{2\pi m^* kT}{h^2}\right)^{3/2} + \ln\frac{1}{n} + \ln 2\right] - kT\ln\frac{N!}{n!(N-n)!} - kT(N-n)\ln 2 \quad , \tag{2.49}$$

where n is the concentration of the free electrons in the band, and m^* is the effective mass. In this expression, the first term nE is similar to the the term $N\tilde{E}$ in (2.11), the second term corresponds to the kinetic energy of the free electrons in the band, and the remaining terms take account of the spin of the electrons. From (2.49) we see that for $n \ll N$ (i.e., when the donors are weakly ionized) these additional terms can be neglected; this is also the case when $N \simeq M$ (in the energy scheme of Fig. 2.1), corresponding to the case of strongly filled traps, which we assumed in Sect. 2.1. In the general case of arbitrarily filled levels, the additional terms must be included in (2.10,11).

Well below the Curie point $F_1 \gg F_2$, and the electrons make no substantial contribution to the free energy. On the other hand, in the paraelectric phase, where $P = 0$, the free energy $F = F_0 + F_2$, and hence the equilibrium configuration of the free electrons, n, is determined from the condition

$$(\partial F/\partial n)_T = (\partial F_2/\partial n)_T = 0 \quad . \tag{2.50}$$

Assuming that $n \ll N$, we obtain from (2.50) the well-known expression for the concentration of the free electrons in the paraelectric phase,

$$n(E_0) = 2(NN_c)^{1/2}\exp(-E_0/2kT) \quad , \tag{2.51}$$

where E_0 is the activation energy of the donor centers in the paraelectric region, and N_c is the density of states in the band,

$$N_c = (2\pi m^* kT/h^2)^{3/2} \quad . \tag{2.52}$$

In the ferroelectric region, near the phase transition, F_2 makes a substantial contribution to the free energy of the crystal, and from the condition $(\partial F/\partial n)_T = 0$ we obtain an expression similar to (2.51) for the concentration of the free electrons,

$$n(E) = 2(NN_c)^{1/2}\exp(-E/2kT) \quad , \tag{2.53}$$

where E is the activation energy of the donors in the ferroelectric region. By analogy with (2.12) and (2.46) the energy E can be expanded as a series of even powers of the polarization,

$$E = E_0 + \frac{1}{2} a_1 P^2 + \frac{1}{4} b_1 P^4 + \frac{1}{6} c_1 P^6 + \dots \quad . \tag{2.54}$$

The solution of (2.53) determines the dependence $n = n(P)$ in the vicinity of the Curie point. Comparing (2.51), (2.53), and (2.54), we conclude that the phase transition in a ferroelectric semiconductor is accompanied by an anomaly in the temperature dependence of the carrier concentration n due to the change in the activation energy of the donors near the Curie point. An analysis of (2.54) similar to the one carried out in Sect. 2.4 shows that for first-order phase transitions there is a jump in activation energy, ΔE, at the phase transition, and for second-order phase transitions the activation energy undergoes a continuous change. These peculiarities, in turn, result in the appearance of anomalies in the temperature dependence of the carrier concentration near the Curie point. Anomalies of the concentration n near the Curie point may also be of a different nature. They may, in particular, be due to the changes in the effective mass of the carriers (and hence the density of states) and the concentration of the donor centers. In principle, in the phase transition region there may appear centers of a different nature, for instance levels associated with the nuclei of a new phase (Känzig regions).

In the general case, the appearance of the nuclei of a new phase or fluctuations due to a local change in the transition parameter can be stabilized by the capture of an electron and production of new quasiparticles (phasons and fluctuons, see Sect.3.6) near the phase transition. The formation of these new electron states, may, in turn, cause an anomaly of the carrier concentration n near the phase transition.

As will be shown in the next chapter, the microscopic nature of the effect of the electron subsystem on a phase transition in ferroelectrics consists in the interaction of soft optical phonons (the soft ferroelectric mode [1.5]) with free or localized electrons. It is therefore natural to assume that the indicated anomaly in the behavior of the activation energy of the donor centers in the vicinity of the Curie point, namely the possible change in the concentration and nature of the centers themselves, is due to the appearance of the bound state of optical phonons and electrons forming quasiparticles [2.20]. In this case the appearance of the bound state only near the Curie point is a manifestation of the well-known energy decay threshold [2.20], which is caused, in the case of a ferroelectric semiconductor, by the temperature dependence of the energy of the transverse optical phonons near the Curie point (the soft mode).

The electrical conductivity σ of a ferroelectric semiconductor is related to the concentration of free carriers n by

$$\sigma = qn\mu \quad , \qquad (2.55)$$

where q is the magnitude of the charge of the electron, and μ is the mobility. Thus the possible anomalies in the temperature dependence $n = n(T)$ cause the peculiarities in the electrical conduction of a ferroelectric semiconductor near the Curie point. From (2.55) we see that another possible cause of electrical conduction anomalies near the Curie point is the behavior of the mobility near the phase transition. This and other possible mechanisms to explain electrical conduction anomalies in ferroelectric semiconductors are considered in [1.7,14].

It is well known that the band structure of a number of ferroelectric semiconductors, for instance those with the perovskite structure [1.5], is characterized by a conduction band with six minima located along the main symmetry axes. In the paraelectric phase these minima are degenerate. Below the phase transition temperature, at $T < T_1$, the degeneracy is lifted because of the tetragonal distortion of the lattice. Then the energy of the minima located in the direction of the spontaneous polarization [001] is lower than the other minima (located along the axes [100] and [010], by an amount $V \equiv V(P,\sigma_k)$. This leads to redistribution of the conduction electrons among the minima.

Suppose that the electrical conduction is of an extrinsic nature and is due to completely ionized donors with a concentration N and an activation energy E. For the tetragonal distribution considered above, we denote the electron concentrations at the minima of the conduction band by $n_x = n_y$, n_z. Then the free energy of the electron subsystem can be written, by analogy with (2.11) [2.6], as

$$F_2 = 2n_x E_g + n_z (E_g - V) - NE \quad . \tag{2.56}$$

Using the redistribution conditions

$$n_x = n_y = n_z \exp(- V/kT) \tag{2.57}$$

and the electrical neutrality condition

$$2n_x + n_z = N \tag{2.58}$$

we obtain the following expression for the free energy:

$$F_2 = N\tilde{E} - N \frac{V}{2\exp(-V/kT)+1} \quad , \tag{2.59}$$

where

$$\tilde{E}(P, \sigma_k) = E_g - E \quad . \tag{2.60}$$

The expression (2.59) for the free energy of the electron subsystem differs from (2.11) by a correction term dependent on the energy V. The correction introduced by the inclusion of the many-valleyed structure of the ferroelectric semiconductor can make a substantial contribution to F_2 only when $V \gg kT$. Indeed, for $V < kT$ we have

$$N \frac{V}{2\exp(-\frac{V}{kT})+1} \simeq N \frac{V}{3}\left[1 + \frac{2}{3}\frac{V}{kT} + \frac{2}{9}\left(\frac{V}{kT}\right)^2\right] \quad , \tag{2.61}$$

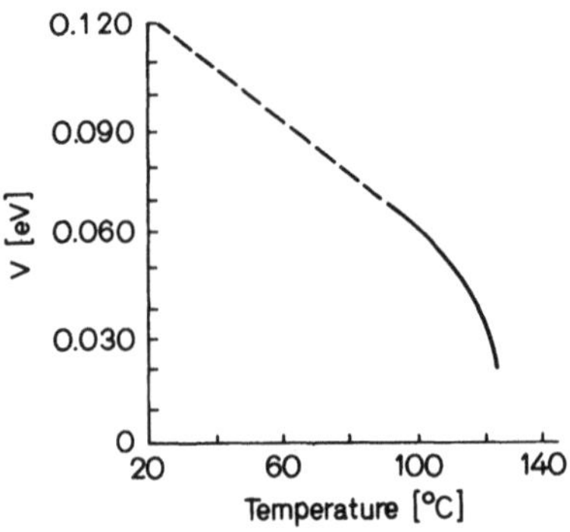

Fig. 2.4. Experimental dependence V(T) for $BaTiO_3$ crystals [2.6]

and since $V \ll \tilde{E}$, the correction is insignificant. When $V \gg kT$, the correction term $\sim NV$, and the inclusion of the many-valleyed structure may prove significant. Figure 2.4 depicts the temperature dependence $V = V(T)$ for $BaTiO_3$ obtained from electrical conductivity measurements [2.6]. These data show that for $BaTiO_3$, near the phase transition $V \simeq 0.03$-0.05 eV, and the correction term is insignificant. It should, however, be borne in mind that the effect of the electrons on the phase transition is determined by the derivatives of the energy V with respect to the polarization, for instance $\partial^2V/\partial P^2$, which may be comparable with the corresponding derivatives of the energy $\tilde{E}$, and hence the inclusion of the many-valleyed structure may be significant. But an equally significant contribution can be made by the traps [see (2.11)] if we take into account the dependence of their energies u_1 and u_2 on the polarization (2.54). If we also recall that in high-resistance ferroelectric semiconductors a contribution to the free energy can be made by several levels or by a whole system of levels, we find that inclusion of the many-valleyed structure is one of many corrections which it is difficult to single out experimentally.

3. The Microscopic Theory of Photoferroelectric Phenomena

It is well known that the mechanism of phase transitions in crystals is associated with instability of atomic vibrations. In particular, the mechanism of ferroelectric phase transitions in contemporary dynamic theory is conditioned by the instability of one of the transverse optical vibrations at the center of the Brillouin zone, and the anharmonicity of the corresponding vibration branch is considered to be responsible for the instability. This mechanism was first suggested by GINSBURG [29] and COCHRAN [3.1], who demonstrated the agreement between this microscopic approach and the thermodynamic theory of ferroelectrics. Before proceeding to the specific features of the dynamic theory of ferroelectric semiconductors we shall consider briefly the pecularities of optical vibrations in a ferroelectric at $\underset{\sim}{k} = 0$ and their relationship with the ferroelectric phase transition.

3.1 Ferroelectric Phase Transitions and the Soft Vibration Mode

Consider the equation of motion of an atomic anharmonic oscillator in an external electric field $\mathscr{E}$,

$$m\ddot{\xi} + r\dot{\xi} + k\xi + s\xi^3 = q\mathscr{E} \quad . \tag{3.1}$$

Here, m and q are the effective mass and charge of the atom, respectively; r, k, and s are coefficients, and ξ is the generalized coordinate of the atom. Considering a set of such oscillators with a concentration N for a variable field $\mathscr{E} = \mathscr{E}_0 \exp(i\omega t)$, the polarization $\tilde{P} = q\xi N$ is determined by

$$\mu\ddot{\tilde{P}} + \nu\dot{\tilde{P}} + \alpha\tilde{P} + \beta\tilde{P}^3 = (\mathscr{E}_0/2)\exp(i\omega t) \quad , \tag{3.2}$$

where

$$\mu = \frac{m}{2q^2N} \quad , \quad \nu = \frac{r}{2q^2N} \quad , \quad \alpha = \frac{k}{2q^2N} \quad , \quad \beta = \frac{s}{2q^4N^3} \quad .$$

If the external field is constant ($\omega = 0$), then (3.2) reduces to the equation of state (2.16) already known from ferroelectric thermodynamics

$$\alpha\tilde{P} + \beta\tilde{P}^3 = \mathscr{E}_0/2 \tag{3.3}$$

Thus the polarization $\tilde{P}$ determined for the anharmonic oscillator system (3.1), can be identified with the spontaneous polarization P of ferroelectrics. Hence it follows, in turn, that the dielectric constant of a ferroelectric can be determined from (3.2) by analogy with (2.26,27). We find

$$\varepsilon = \varepsilon_\infty + \frac{2\pi}{\alpha - \mu\omega^2 + i\nu\omega} \qquad \text{for } T > T_0 \quad ,$$

$$\varepsilon = \varepsilon_\infty + \frac{\pi}{-\alpha - \mu\omega^2 + i\nu\omega} \qquad \text{for } T < T_0 \tag{3.4}$$

and α satisfies the temperature dependence (2.22).

Comparing (3.4) with the frequency dispersion of the dielectric constant,

$$\varepsilon = \varepsilon_\infty + \frac{\varepsilon_0 - \varepsilon_\infty}{1 - \left(\frac{\omega}{\omega_{TO}}\right)^2 + i\gamma\left(\frac{\omega}{\omega_{TO}}\right)} \quad , \tag{3.5}$$

we obtain the following temperature dependence of the frequency of the transverse optical vibration, ω_{TO}, for $T > T_0$:

$$\omega_{TO}^2 = \frac{\alpha}{\mu} = \frac{\alpha_T'}{\mu}(T - T_0) \quad . \tag{3.6}$$

Thus, GINSBURG's model of the anharmonic oscillator suggests the existence, near the Curie point, of an optical vibration, whose frequency depends anomalously on the temperature and vanishes at the Curie point. The corresponding vibration branch is termed the ferroactive, or 'soft', vibration mode in the literature. Since the dispersion relations (3.4,5) used above are valid for long wavelength $\lambda \gg a$ (where a is the lattice constant), (3.6) corresponds to the center of the Brillouin zone, $\underline{k} = 0$. Note that the shift of the Curie point with a change in electron concentration can be obtained from (3.6) and corresponds to the change in the elasticity coefficient of the oscillator, α, in this model.

The soft vibration mode and the corresponding ferroelectric phase transition of the displacement type also emerge from the general theory of anharmonic vibrations of crystal lattice atoms [3.1]. As is well known, in the harmonic approximation the dispersion of the lattice vibration frequencies [i.e. the dependence $\omega_i = \omega_i(\underline{k})$, where $\underline{k}$ is the wave vector ($|\underline{k}| = 2\pi/\lambda$)] can be obtained from the condition of vanishing of the determinant

$$|B_{\alpha\beta}^{kk'} - m_{k'}\,\omega^2\delta_{kk'}\,\delta_{\alpha\beta}| = 0 \quad , \tag{3.7}$$

where $B_{\alpha\beta}^{kk'}$ are coefficients depending on $\underline{k}$ and the force constants; $m_{k'}$ are the atomic masses, $k, k' = 1,2,\dots,s$; $\alpha,\beta = x,y,z$. If the unit cell of the lattice contains s different atoms, the characteristic equation (3.7) of degree 3s has, in the general case, 3s real solutions $\omega_i = \omega_i(\underline{k})$ $(i = 1,2,\dots 3s)$ or 3s vibration branches. The characteristic equation (3.7) becomes simpler if we consider a diatomic lattice and make full use of its symmetry. In this case the problem amounts to determining the dispersion of the vibration frequency of a linear atomic chain characterized by three vibration branches, one longitudinal and two transverse.

To obtain the soft transverse optical mode in the long-wavelength approximation ($\underline{k} = 0$) COCHRAN [3.1] considered a special model of atomic interaction in a diatomic crystal. The forces of interaction between adjacent atoms in this model can be divided into Coulomb (long-range) forces, which depend on the polarization P, and non-Coulomb (short-range) forces. The non-Coulomb interaction forces determine the vibration anharmonicity in this model.

The final expression for the frequency of the transverse optical vibration at $\underline{k} = 0$ is

$$\mu\omega_{TO}^2 = R_0 - \frac{4\pi(\varepsilon_\infty+2)(Zq)^2}{9v} \tag{3.8}$$

where m_1, m_2 are the masses of two adjacent interacting ions;

$$\mu = \frac{m_1 m_2}{m_1 + m_2}$$

is the reduced mass; Z is the effective charge; v is the volume of the unit cell; and R_0 is the force constant of the short-range non-Coulomb interaction.

The frequency of the longitudinal optical vibration ω_{LO} at $\underline{k} = 0$ can be similarly obtained:

$$\mu\omega_{LO}^2 = R_0 + \frac{8\pi(\varepsilon_\infty+2)(Zq)^2}{9v\varepsilon_\infty} \tag{3.9}$$

The relations (3.8,9) must be supplemented by the Lyddane-Sachs-Teller relation

$$\frac{\omega_{LO}^2}{\omega_{TO}^2} = \frac{\varepsilon_0}{\varepsilon_\infty} \tag{3.10}$$

and the relationship between the force constant R_0 and the compressibility χ,

$$R_0 = 6\frac{r_0}{\chi} \quad , \tag{3.11}$$

where r_0 is the distance between the nearest atoms in a simple structure of the type of NaCl.

A comparison of (3.8) and (3.10) shows that the Curie-Weiss law holds, provided that

$$R_0 - \frac{4\pi(\varepsilon_\infty+2)(Ze)^2}{9v} \sim (T - T_0) \quad . \tag{3.12}$$

This, in turn, indicates that in COCHRAN's model the frequency of the transverse optical vibration at the Curie point reduces to zero and, hence, the phase transition and the dielectric anomalies are due to the existence of the soft vibration mode. An analysis of the left-hand side of (3.12) shows that the temperature dependence of v and ε_∞ can be neglected, and therefore the temperature dependence of ω_{TO}^2 is due to the strong dependence of the force constant R_0 on the amplitude of the ion shift. The latter fact is the evidence of a substantial vibrational anharmonicity. From (3.9), we see that the frequency of the longitudinal optical vibration does not vanish at the Curie point. Thus, for a phase transition, in particular a ferroelectric one, it is only necessary that one of the vibration modes be unstable.

3.2 Effect of Screening on the Soft Vibration Mode

If a ferroelectric semiconductor has a sufficiently high concentration of free carriers, the Coulomb interaction of the lattice ions can be screened. In terms of Cochran's theory this reduces the Lorentz field (as well as the microscopic field) and shifts the temperature at which the frequency of the transverse optical vibration mode vanishes (the Curie-point shift). The temperature dependence ω_{TO} taking screening into account was calculated in [3.2]. It is important to note that both in [3.2] and in other investigations dealing with the effect of screening on the Curie point [3.3-5] the electrons do not affect the vibration anharmonicity and only screen the Coulomb ion-ion interaction.

If screening is included in Cochran's model for an ion lattice with a structure of NaCl, the relation (3.8) reduces to the form

$$\mu\omega_{TO}^2 = R_0 - \frac{4\pi(\varepsilon_\infty+2)(Zq)^2}{9v} G(\lambda\tfrac{a}{2}) \quad , \tag{3.13}$$

where a is the lattice constant; G(x) is the correlation function

$$G(x) = \frac{1}{2\pi}\sum \frac{x^2}{(n^2+m^2+p^2)^{1/2}} \exp\left[-x(n^2+m^2+p^2)^{1/2}\right] \quad , \tag{3.14}$$

where the summation extends over all triples (n,m,p) such that n + m + p is odd; and λ is the screening constant, satisfying

$$\ell_D = \lambda^{-1} = \left(\frac{\varepsilon kT}{8\pi q^2 n}\right)^{1/2} \tag{3.15}$$

where ℓ_D is the Debye length, and n is the concentration of free electrons. Since $G(x) = 1$ at $x = 0$ in the absence of electrons ($\lambda = 0$) the relation (3.13) reduces to (3.8).

At the Curie point $\omega_{TO}^2 = 0$ (the soft mode). The Curie-point shift due to the electrons can be obtained from the relations

$$R_0 - \frac{4\pi(\varepsilon_\infty+2)(Zq)^2}{9v} = 0 \qquad \text{for } T = T_0 \quad , \tag{3.16}$$

$$R_0 - \frac{4\pi(\varepsilon_\infty+2)(Zq)^2}{9v} G(\lambda\tfrac{a}{2}) = 0 \qquad \text{for } T = T_0' \quad , \tag{3.17}$$

and the temperature dependence of $G(\lambda\frac{a}{2})$ due to the fact that λ and a are functions of the temperature.

To calculate T_0' from (3.16,17) it is necessary to determine the force constant R_0 as a function of the temperature. Assuming (as was emphasized above) that the electrons do not affect R_0, we determine the temperature dependence $R_0 = R_0(T)$ from the relations (3.8-10) and from the Curie-Weiss law in the form

$$\varepsilon_0 = \varepsilon_\infty + \frac{C}{T-T_0} \quad . \tag{3.18}$$

For $\varepsilon_\infty \simeq 1$ these calculations lead to the following dependence $R_0 = R_0(T)$:

$$R_0 = \frac{4\pi(Zq)^2}{3v}\left[\frac{3}{C}(T - T_0) + 1\right] \quad . \tag{3.19}$$

If we substitute (3.19) into (3.17), then, neglecting the temperature dependence of the unit cell volume v, we get

$$T_0' = T_0 - \frac{C}{3}\left[1 - G(\lambda\tfrac{a}{2})\Big|_{T=T_0'}\right] \quad . \tag{3.20}$$

From (3.20) it is seen that since $dG/d\lambda < 0$, an increase in the concentration of free electrons n shifts the Curie point toward the low temperatures. This was the result (2.36) we earlier obtained from thermodynamics.

The radical difference between (3.20) and (2.36) is that in (2.36) N stands for the total concentration of carriers, both free and localized in deep traps, whereas in (3.20) the screening constant λ depends by (3.15) only on the concentration of free carriers n. For shallow traps, which are in thermal equilibrium with the corresponding band, this difference is found to be insignificant, and λ depends on N [3.6]. However, when $u/kT \gg 1$, i.e. for sufficiently deep traps, $\lambda = \lambda(n)$ and hence the physical meaning of (2.36) and (3.20) is different. This difference becomes particularly clear if we consider photoferroelectric phenomena, for instance the Curie-point shift by nonequilibrium carriers.

The Curie point shift described by (3.20) results from the screening of the internal fields in a ferroelectric by free nonequilibrium carriers. The physical significance of (2.36) consists in the effect on the Curie temperature of optical recharging of the levels in the ferroelectric. One of the possible mechanisms for this effect will be discussed in the next section. Here we only give some quantitative estimates. Table 3.1 lists numerical values of the function G(x) for $0.1 \leq x \leq 1$. Using these values and (3.15,20), we find that at a free-carrier concentration of $n = 10^{17} - 10^{18}\ cm^{-3}$ and $C = 10^4 - 10^5$, the Curie point shift $\Delta T_0 = T_0 - T_0' = 1 - 10$ K. Estimation by (2.36) with an allowance for (2.47) shows that a Curie-point shift of $\Delta T_0 = 1 - 10$ K is achieved at concentrations N of the same order as that of n above (see Chap.5). Such a high concentration of free nonequilibrium carriers in a ferroelectric is clearly impossible, and n and N must signify, in both cases, the concentration of the recharged levels (traps) in the crystal. As noted above, the sign of the effect (the direction of the Curie-point shift) is the same for screening and for optical-level recharging. Therefore experimental separation of these two mechanisms requires conditions in which it is possible to distinguish the localized and free carriers (see Chap.5).

An attempt was made [3.5] to take into account within the framework of the screening mechanism both the free carriers and those captured by deep traps, and to separate, as far as possible, their effect on the phase transition temperature. The approximation used was that the Debye length (3.15) depends on

Table 3.1. Numeral values of the function G(x)

x	G(x)	x	G(x)
0.1	0.99913	0.6	0.96937
0.2	0.99654	0.7	0.95859
0.3	0.99223	0.8	0.94633
0.4	0.98624	0.9	0.93268
0.5	0.97860	1.0	0.91770

the dielectric constant ε, and that the dielectric constant of the semiconductor, in turn, depends on the concentration of free and localized carriers. Indeed, near the center of the Brillouin zone (for small values of $|\underline{k}| = 2\pi/\lambda$) [3.7]

$$\varepsilon(k) = \varepsilon_\infty + \delta + \frac{\lambda_0^2}{|\underline{k}|^2} , \tag{3.21}$$

where

$$\delta = \frac{4\pi\hbar^2 q^2}{u^2 m^*} N , \quad \lambda_0^2 = \frac{8\pi q^2 n}{kT} . \tag{3.22}$$

Here, ε_∞ is the electron dielectric constant in the absence of carriers ($n = N = 0$), n is the carrier concentration in the zone, N is the carrier concentration at levels with an activation energy u, and m^* is the effective mass of the electrons (holes) localized at the levels. If we further assume that even in the presence of localized electrons the screening of the ion-ion interaction is not associated with the anharmonicity and does not affect it, the localized carriers can be included by substituting the screening constant λ into the expression (3.13) for the frequency of the transverse optical phonon taking (3.21,22) into account,

$$\lambda = \lambda_0/(\varepsilon_\infty + \delta)^{1/2} \tag{3.23}$$

Substituting (3.23) into (3.13) yields the following expression for the Curie-point shift:

$$T_0' = T_0 - \frac{C}{3\varepsilon_\infty}\left[1 - \left(1 - \frac{\delta}{\varepsilon_\infty}\right)G\left(\lambda\frac{a}{2}\right)\Big|_{T=T_0'}\right] . \tag{3.24}$$

This result makes in possible to analyze separately the effect of the free and localized carriers on the Curie point.

In the absence of localized carriers $N = 0$, and for $\varepsilon_\infty = 1$, (3.24) reduces to (3.20). Consider now the opposite case, when the concentration of the free carriers can be neglected as compared with that of the localized ones: $n \ll N$.

This is the case in high-resistance ferroelectric semiconductors. Here, (3.24) becomes

$$T_0' = T_0 - 4\pi\left(\frac{\hbar q^2}{u\varepsilon_\infty}\right)^2 \frac{C}{3m^*} N \quad . \tag{3.25}$$

This shows clearly the linear dependence of the Curie-point shift on the concentration of localized electrons N, which agrees with the linear dependence (2.36) following from thermodynamics. Substituting $N = 10^{18}$ cm^{-3}, $u = 0.5$ eV, $C = 10^5$ K, and $\varepsilon_\infty = 10$ (see Chap.5) into (3.25), we get $|\Delta T_0| = T_0 - T_0' = 9m/m^*$, where m is the mass of a free electron or polaron. For localized carriers $m/m^* < 1$. In any case, (3.25) agrees qualitatively with the experimental data, which will be discussed in more detail in Chap. 5. Nevertheless, quantitative distinction between the free and localized carriers on the basis of (3.24) is hardly possible, especially because of the difficulties involved in estimating the parameter m/m^*.

3.3 Ferroelectric Phase Transitions and Interband Electron-Phonon Interactions

The screening mechanism and the participation of the electron subsystem of a crystal in the phase transition, which were considered in the preceding section, were associated with the change in Coulomb ion-ion interaction and did not affect the short-range forces, which cause the vibration anharmonicity in Cochran's model. This resulted in renormalization of the frequency of the transverse optical phonon and hence in a Curie-point shift at a constant value of the Curie-Weiss constant. At the same time it is clear that the influence of the electrons may not amount to such a simple effect and may have a direct bearing on the anharmonicity mechanism responsible for the phase transition. It is impossible to include the influence of the electrons on the anharmonicity of the ferroelectric mode of vibration within the framework of Cochran's theory, since in this case the cause for the anharmonicity remains undisclosed. It has been shown that the electron-phonon interaction affects the temperature dependence of the ferroelectric soft mode in oxide perovskites [3.8,9].

The microscopic mechanism responsible for the direct influence of the electrons on the phase transition, at least for ferroelectric phase transitions, is the interband electron-phonon interaction [3.10-12]. Since we have already discussed the main features of the theory of ferroelectric phase transitions, both on the basis of lattice dynamics and on the basis of the Landau-Ginsburg phenomenological theory, we shall only discuss briefly the role of interband electron-phonon interactions in the phase transition mechanism. As a corollary we shall consider some effects associated with the influence of electron excitations on phase transitions.

In Sect. 3.1 it was shown that the anharmonicity of one of the transverse optical vibrations at $\underline{k} = 0$, and hence its instability, is the cause of the ferroelectric phase transition. One of the possible mechanisms responsible for the instability of vibrations, and therefore also for the phase transition, is the interband electron-phonon interaction. This mechanism consists in the interaction of the electrons of two adjacent bands of a crystal with one of the optical vibrations, one of these bands being empty (or nearly empty), and the other being completely filled with electrons. Such interaction, or 'intermixing' of adjacent energy bands, results, on the one hand, in a change of the frequency of the interacting optical phonon and, on the other, in a change of the electron spectrum (the energy gap). This interaction of a filled and an empty band leads to instability of the intermixing optical vibration and, correspondingly, to a transition from a symmetric to a less symmetric phase. This phase transition was termed the 'pseudo-Jahn-Teller effect' by analogy with the usual Jahn-Teller effect which describes the reduction of symmetry of a molecule due to the interaction of a degenerate and a nondegenerate electron level [3.11,12].

This model gives the following expression for the temperature T_1 of the phase transition [3.12]:

$$kT_1 = \frac{E_{g0}}{4} (\tanh^{-1}\tau)^{-1} \tag{3.26}$$

where

$$\tau = \frac{2V^2}{M\omega^2 E_{g0}} \tag{3.27}$$

Here V is the constant of the interband electron-phonon interaction, M is the corresponding mass factor, ω is the frequency of the active vibration mode, and E_{g0} is the energy gap. A necessary condition for the phase transition is [3.12]

$$\tau > 1 \qquad (3.28)$$

Thus, τ is a key parameter in the theory of interband electron-phonon interactions. The condition (3.28) means that with given values of $M\omega^2$ and of the energy gap E_{g0} the phase transition only takes place for sufficiently strong interband interaction.

The model considered above makes it possible to explain the entire range of photoferroelectric phenomena. As an example consider a photoconducting crystal which has high symmetry above the phase-transition temperature T_1. It follows from thermodynamics that illumination of the crystal in the spectral region where it exhibits photoconduction lowers the phase transition temperature (see Sect.2.3). Suppose the illumination excites the electrons from the lower to the upper energy band, so that the electron concentration in the upper band increases by Δn. Then, according to (3.26), the phase-transition temperature shifts toward the lower temperatures by the amount

$$\Delta T = \frac{E_g}{4k}\left[\left(\tanh^{-1}\frac{\tau}{1+2\Delta n\tau}\right)^{-1} - (\tanh^{-1}\tau)^{-1}\right] \quad . \qquad (3.29)$$

The physical meaning of this effect is that photoactive light absorption changes the filling of the bands with electrons, and hence also changes the contribution to the free energy due to the interband electron-phonon interaction. As indicated above, photoactive illumination changes not only the concentration of free electrons in the band, but also the filling of all the traps in the forbidden band which are due to impurities or defects. Here (and this has also been emphasized), for high-resistance ferroelectric semiconductors a shift in phase transition temperature must also be of an 'extrinsic' nature, since the concentration of electrons at the levels and its change upon illumination may be a few orders higher than in the band. In contrast to the mechanism based on screening (Sect.3.2), there is no radical difference between the participation of free and localized carriers. The only additional criterion for this mechanism to apply to localized electrons is the condition (3.28), which imposes a restriction on the activation energy of the corresponding levels at a given value of the constant of the electron-phonon interaction.

Equation (3.26) makes it possible to relate the Curie point of a ferroelectric, T_1, with the energy gap E_{g0}. The qualitative results agree with experiment. For instance, a comparison of the Curie point of the ferroelectric semiconductor GeTe, which has a narrow energy gap, with that of dielectric $BaTiO_3$, which has a broad energy gap, confirms (3.26).

The condition (3.28) imposes a restriction on E_{g0}: at fixed values of the electron-phonon interaction constant V and, accordingly, of the vibration anharmonicity, determined by the frequency ω, the energy gap E_{g0} must not be too large. The physical meaning of this restriction is as follows. For ferroelectrics with a sufficiently narrow energy gap the intrinsic vibration anharmonicity is inessential. The phase transition from the high-symmetry paraelectric phase to the low-symmetry ferroelectric phase is associated with the anharmonicity of the optical vibration mode, which intermixes the bands. In this case the temperature dependences of the frequency of the soft mode, the spontaneous polarization, the dielectric constant, and other macroscopic parameters are determined by the filling of the adjacent bands [3.13]. Suppose now that we have a ferroelectric with a broad energy gap for which the parameter E_{g0}/kT is so large that the thermal transitions of electrons from the lower to the upper band can be neglected for the entire temperature range of interest, including the phase transition temperature. In this case the anharmonicity of the optical vibration at $\underline{k} = 0$, although associated with interband electron-phonon interaction, can actually be regarded as the intrinsic anharmonicity (by analogy with the intrinsic anharmonicity in COCHRAN's theory [3.1] or in VAKS's theory [1.6]).

In this connection we wish to stress once again the possible role of the local levels in the energy gap of a ferroelectric. It is not improbable that in broad-band ferroelectrics the phase transition is associated with band-level interaction, and E_{g0} should simplify the activation energy of the corresponding levels. For such an extrinsic ferroelectric the Curie point and the temperature dependences of the parameters are determined by the filling of the levels and the band nearest to them.

Along with the intermixing of adjacent energy bands by active optical vibration, a substantial role may be played in an extrinsic ferroelectric by the mechanism of interaction of two (or more) local levels (centers) with a phonon corresponding to the local vibrations of the centers. This mechanism was developed in detail by CHANUSSOT [3.13] for the case of barium titanate with an iron impurity.

At sufficiently small concentrations, iron is incorporated in the lattice of a perovskite ferroelectric in such a way that the charge of the two Fe^+ ions is compensated by the one oxygen vacancy. One ion enters one oxygen octhedron, forming an FeO_6 complex, while the other ion substitutes for a

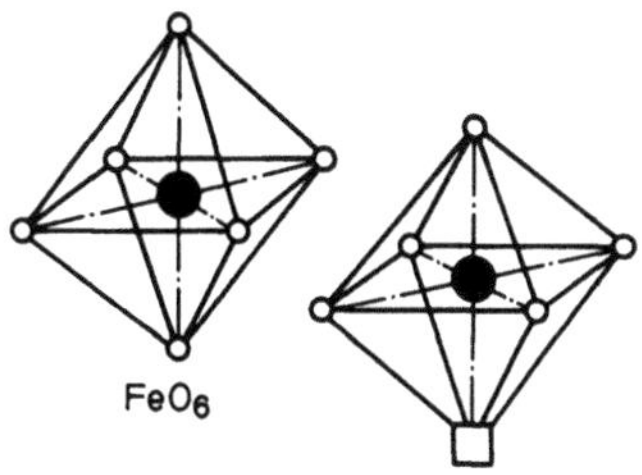

Fig. 3.1. Iron impurity centers in $BaTiO_3$

titanium atom in the other oxygen octahedron, forming an FeO_5 complex plus an oxygen vacancy (see Fig.3.1). In the forbidden band of $BaTiO_3$, the donor and acceptor levels correspond, respectively, to these impurity centers. Transition of an electron from the donor level of FeO_6 to the acceptor level of FeO_5 and interaction of the recharged centers with one of the local vibration modes results, due to the pseudo-Jahn-Teller effect, in instability of the corresponding mode and in a change of the symmetry of the impurity complexes. The symmetry of the centers of FeO_6 and FeO_5 + oxygen vacancy changes correspondingly ($O_h \rightarrow C_{4v}$ and $C_{4v} \rightarrow C_s$). This, in turn, leads to cell deformation and makes a contribution to the spontaneous polarization of the crystal. Thus, along with the 'intrinsic' Jahn-Teller effect associated with interband electron transitions, in ferroelectrics one may observe an 'extrinsic' Jahn-Teller effect. The latter may be responsible for the influence of the impurity, and its optical recharging as a result of the extrinsic photoeffect, on the phase transition and the ferroelectric properties near the Curie point. Thus, this mechanism has been used to explain some photoferroelectric phenomena in barium titanate with an iron impurity, such as the influence of photoactive illumination on the domain structure, spreading of the phase transition, etc. [3.13]. We shall discuss this in more detail in Chaps. 5, 8.

3.4 Change in Dipole Moment of Centers on Optical Recharging

We now turn to another possible microscopic mechanism of the influence of nonequilibrium carriers on the spontaneous polarization, a mechanism based on the change in the dipole moment of the impurity centers as a result of optical recharging [3.14]. In this case, the change in spontaneous polarization

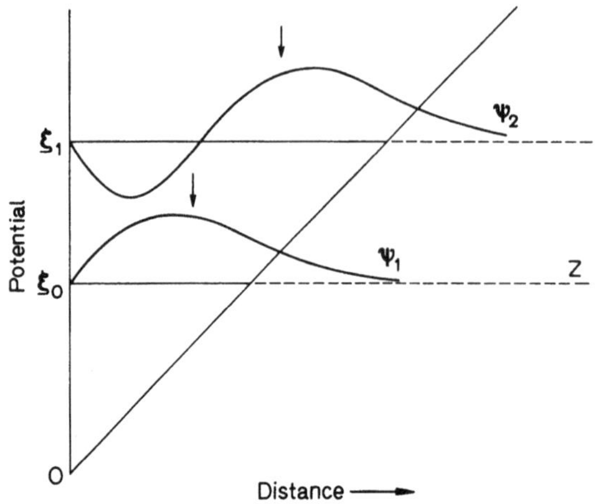

Fig. 3.2. Eigenfunctions and eigenstates for a triangular potential barrier [3.21]

ΔP_0 is related linearly to the concentration of the recharged centers, N,

$$\Delta P_0 = dN = \alpha^* f P_0 N \quad , \tag{3.30}$$

where $d = \alpha^* f P_0$ is the change in the dipole moment of the center, α^* is the impurity polarizability, f is the Lorentz factor, and P_0 is the spontaneous polarization. Why the dipole moment of the center changes when it is on recharged or excited is easily seen from Fig. 3.2 [3.21]. Let us consider the transition of an electron from the ground level ξ_0 to the localized excitation level ξ_1 in the presence of an asymmetric potential, of say, triangular shape. The solution of the Schrödinger equation for this potential is given by the wave functions $\psi_i(x)$ presented in Fig. 3.2. The arrows show the average position of the electron, determined by $\psi_i|x|\psi_i^*/\psi_i\psi_i^*$, for the ground and excited states. Thus, the transition of an electron between two localized states in the field of an asymmetric potential results in its displacement in the coordinate space. Since by symmetry this displacement has the same direction for all the centers in a pyroelectric, the excitation of the centers changes the microscopic polarization.

The change in the dipole moment of the centers on optical recharging, in turn changes the lattice polarization, since the recharged center polarizes its environment. This makes an additional contribution to the change in the macroscopic polarization of the ferroelectric, ΔP_0. To calculate this contribution LEVANYUK and OSIPOV [3.14] proceeded from the expression for the free energy that takes into account the nonuniform spatial distribution of the spontaneous polarization P_0. For the case of deep impurity centers, this expression is

$$F = \frac{d^2}{2\alpha^*} + \frac{1}{4}\beta^* d^4 - fd(P_0 + \Delta P_{01}) + \frac{1}{2} a(P_0 + \Delta P_{01})^2$$
$$+ \int \left[\frac{1}{2}\alpha P^2 + \frac{1}{4}\beta P^4 + \frac{1}{2}\kappa (\nabla \underline{P})^2\right] dr \qquad (3.31)$$
$$+ \frac{1}{2}\int\int \frac{\mathrm{div}\underline{P}(r)\,\mathrm{div}\underline{P}(r')}{|\underline{r}-\underline{r}'|}\, d\underline{r}\, d\underline{r}' \quad .$$

Here, ΔP_{01} is the change in polarization due to interaction of the impurity with the lattice. From (3.31) LEVANYUK and OSIPOV obtained for the concentration of the recharged centers, N,

$$\Delta P_{01} = \frac{(\alpha^* f^2 - a)cP_0}{4\pi(c+a)} N \quad , \qquad (3.32)$$

where $c = 16(\kappa/\pi)^{1/2} r^2$, r is the radius of the impurity ion, κ is the correlation constant, and a is the constant of the deformation potential, which was introduced in (2.13).

The total change in spontaneous polarization is thus determined by the sum of (3.30) and (3.32),

$$\Delta P = \Delta P_0 + \Delta P_{01} = \frac{(\alpha^* f^2 - a)}{4\pi(c+a)} cP_0 N + \alpha^* f P_0 N \quad . \qquad (3.33)$$

The case of shallow impurity centers leads to the same expression with f = 0 [3.14]:

$$\Delta P \simeq -\frac{a}{4\pi} P_0 N \quad , \qquad (3.34)$$

which coincides with (2.37) for b = c = 0. A comparison of (3.33) with (3.34) shows that optical recharging of deep and shallow impurity centers can make contributions of opposite signs to the change in spontaneous polarization, and that (as has already been noted) the free-energy expression (2.10) is not valid for deep impurity centers.

3.5 Phasons and Fluctuons

Ferroelectrics and pyroelectrics are among the many materials in which photostimulated phase transitions are observed. We begin this section with a discussion of a model [3.15-17] of the influence of electrons on first- and second-order phase transitions based on the idea that a fluctuation is stabilized by the localization of electrons on it and by the production of new quasiparticles, phasons and fluctuons.

In first-order phase transitions, the heterophase fluctuation (the nucleus of the β phase appearing in the α phase, which is at equilibrium at a given temperature) increases the free energy of the system due to the difference between the densities of the free energies of the two phases, φ, and the interphase energy. When electrons cross from the conduction band of the α phase to the heterophase fluctuation level and become localized at this level, the electron energy is reduced by the value of the activation energy of this level, u. If this decrease in the free energy of the electron subsystem offsets the increase in free energy due to the formation of the nucleus of the β phase, such a nucleus (phason) becomes stable. If, further, the self-consistent phason radius R_0 exceeds the critical radius of the nuclei of the β phase at a given temperature (and hence at a given value of the difference between the densities of the free energies of the two phases φ) and at a given value of the density of the surface energy of the interface, σ, then a phase transition $\alpha \rightarrow \beta$ takes place.

As is shown in [3.16], if the free energy of the system depends nonlinearly on the corresponding internal parameter (on the spontaneous polarization in the particular case of the ferroelectrics) and, if also $u/kT \gg 1$, then R_0 greatly exceeds the lattice constant and, accordingly, the phason is a macroscopic quasiparticle. This discussion is based on the idea that the free energy of the electrons makes a substantial contribution to the free energy of the lattice near the phase transition temperature (see Chap.2, also [2.5]). The phason model makes it possible to predict, in the most general case, the existence of two groups of phenomena.

Firstly, this model indicates the influence of the electrons, nonequilibrium ones in particular, on a first-order phase transition. Thus, the formation of a phason is greatly facilitated by an increase in the number of electrons localized on it. In [3.16] it is shown that $R_0 \sim n^{2/3}$, where n is the number of localized electrons. Thus an increase in the concentration of nonequilibrium electrons in the conduction band of the α phase promotes the formation of nuclei of critical size at higher values of φ , i.e. it leads to a photostimulated shift in the temperature of the phase transition $\alpha \to \beta$. The same mechanism reduces the temperature hysteresis (the photohysteresis effect).

Secondly, the phason model predicts a change in the electronic properties of the crystal in the phase transition region. According to this model, in the vicinity of the phase transition temperature the crystal acquires new localized electron states corresponding to the energy level of the phason in the forbidden band of the α phase. The new levels may lead to an anomaly in the electroconductivity and photoconductivity and strongly affect the kinetics of the recombination and trapping of nonequilibrium electrons at temperatures near the phase transition. In other cases phasons can be regarded as a nondegenerate gas of mobile quasiparticles with an effective mass M, which makes a contribution to transport phenomena. In all cases the lower the latent heat of the phase transition, the wider is the range of temperatures at which phasons can exist.

Following [3.16], let us estimate the phason concentration near a first-order nonferroelectric phase transition. We consider the region of the α phase containing a spherical nucleus of the β phase which has captured one electron. The free energy ΔF of this region has the form

$$\Delta F(R) = \frac{4\pi}{3} R^3 \varphi + 4\pi R^2 \sigma + E(R) \quad , \tag{3.35}$$

where E(R) is the energy of the ground state of the electron in the spherical potential well (m is the mass of the free electron) given by

$$E(R) = -u + \frac{\hbar^2 k^2}{2m} \quad , \quad k \cot kR = -\left(\frac{2mu}{\hbar^2} - k^2\right)^{1/2} \quad . \tag{3.36}$$

The equilibrium self-consistent values of the phason radius, $R = R_0$, and the activation energy u_0 can be determined from (3.35,36) and from the conditions $\Delta F(R_0) = 0$ and $\partial \Delta F/\partial R = 0$. The following values are obtained [3.16]:

$$\begin{aligned} u_0 &= 11.5\hbar(\sigma/m)^{1/2} \quad , \quad R_0 = 0.6(\hbar^2/m\sigma)^{1/4} && \text{for } \varphi R_0 \ll \sigma \quad , \\ u_0 &= 5.81(\hbar^2/m\varphi)^{3/5}\varphi \quad , \quad R_0 = 0.73(\hbar^2/m\varphi)^{1/5} && \text{for } \varphi R_0 \gg \sigma \quad , \end{aligned} \tag{3.37}$$

$\hbar$ is Planck's constant divided by 2π.

Numerical estimates based on (3.37) give, for $\sigma \simeq 20$ erg.cm^{-2} and $\varphi = 0$, $u \simeq 1$ eV and $2R_0 \simeq 10$ Å. If $\sigma \ll 10^2$ erg·cm^{-2}, then, away from the phase transition temperature (e.g. for $\varphi \simeq 10^9$ erg·cm^{-3}), the order of magnitude of u_0 and $2R_0$ remains the same. Thus, the estimates or R_0 show the phason is a macroscopic quasiparticle. Corrections for the effective mass of the electron lead to an increase in the equilibrium values of R_0.

These results were qualitatively generalized in [3.16] to the case where the β phase is pyroelectric. Then the equilibrium shape of the phason is an ellipsoid extended in the direction of spontaneous polarization, and in the energy expression (3.35) we must take into account the energy of the electron in the spontaneous polarization field.

The model of localized and mobile quasiparticles can be used to estimate the phason concentration [3.16]. In the nondegenerate-gas approximation, the free energy of N_p mobile phasons has the form

$$F_{mob} = N_p \,\Delta F(R) - N_p kT\ln\left[e\,\frac{N}{N_p}\,v\left(\frac{MkT}{2\pi\hbar^2}\right)^{3/2}\right] \quad . \tag{3.38}$$

Here, v is the volume of the unit cell of the α phase, and N is the number of cells. The free energy of N_p localized phason includes the configurational contribution associated with permutation of N_p phasons among N cells of the α phase,

$$F_{loc} = N_p \,\Delta E(R) - kT\ln\left(e\,\frac{N}{N_p}\right) \quad . \tag{3.39}$$

For a nondegenerate electron gas in the α phase we have, according to (2.49),

$$F = -N_e\, kT\ln\left[e\,\frac{N}{N_e}\,v\left(\frac{mkT}{2\pi\hbar^2}\right)^{3/2}\right] \quad . \tag{3.40}$$

Here, N_e is the concentration of free electrons in the α phase. Since the formation of phasons must not affect the free energy at equilibrium, we equate (3.40) to (3.38) or (3.39), respectively, and obtain the following expression for the phason concentration:

$$N_p = N_e\, \delta\exp(-\Delta F/kT) \quad , \tag{3.41}$$

where $\delta = (M/m)^{3/2}$ for quasiparticles and

$$\delta = \left(\frac{2\pi h}{V^{2/3}mkT}\right)^{3/2}$$

for localized phasons ($\delta \gg 1$). For the above numerical parameters $\Delta F(R_0) \leq 10kT$, and therefore, according to (3.41), in the region of the phase transition $\alpha \rightarrow \beta$ a large number of electrons shift to the phason level or to the mobile phason state. Thus, the phason model assumes a substantial effect of the first-order phase transition on the electroconductivity and even a change in the nature of electroconductivity in the phase transition region.

According to the definition given in [3.16] the fluctuon is a macroscopic quasiparticle formed in the region of order-disorder phase transitions by the fluctuation of an order parameter (magnetization, for instance) and its stabilization on electron capture. The fluctuon model also leads to photostimulated phase transitions and hence to the influence of the phase transition on the electronic properties of the material.

The possibility of photostimulated phase transitions was predicted in 1966 in independent investigations [2.5,3.18]. Not only was the substantial role played by the electron subsystem in the thermodynamics of phase transitions established, but also possible micromechanisms of photostimulated phase transitions for specific systems were proposed, in particular, for ferroelectric phase transitions [2.5]. As indicated above, the phason and fluctuon models [3.16] make it possible to consider, from a unified standpoint, the effect of light on phase equilibria, including photosensitive phase transitions in a solid, the phenomena of photoassociation and photocondensation, photostimulated nucleation, and some other related phenomena. Therefore, to conclude this chapter, we shall briefly survey these phenomena, which represent photostimulated phase transitions in nonferroelectric materials.

The phenomenon of vapour condensation in a light beam was described by TYNDALL as far back as 1869 [3.19]. The author questioned the possibility of explaining it on the basis of the assumption of some irreversible reactions occurring in the system. Later on, WILSON discovered the possibility of formation of neutral centers of condensation of water vapours on illumination of the vapours in the near-ultraviolet region [3.20]. In this connection WILSON assumed that the formation of mist drops in a light beam is due to a reduced surface tension of the water as a result of a photochemical reaction of the formation of traces of hydrogen peroxide. This point of view on the mechanism of photocondensation of water vapours was developed in [3.22]. In later investigations into the photocondensation phenomenon, the authors confirmed the decisive effect of traces of foreign substances (ammonia, hydrogen sulphide, sulphur dioxide) and thus demonstrated the possibility of optical sensitization of the process. The fact of sensitization of the effect by ammonia molecules led some authors to believe that photocondensation of water vapours is associated with electron excitation of the ammonia molecules. An increase in dipole moment on excitation results in association of water molecules with those of ammonia. Collisions of the associates produce mist drops. The possibility of formation of neutral condensation centers in the vapours of various liquids under the effect of photoactinic radiation was confirmed in [3.18], where the formation of mist drops in the vapours of benzene, bromobenzene, carbon tetrachloride, and other substances was observed. No traces of photochemical reactions in the system were detected, which in turn suggested that the common cause for photocondensation is an increase in the energy of molecular interaction on electron excitation of the molecules. Thus it was reliably established that on illumination of the vapours of many substances in the region of their intrinsic absorption or in the region of absorption of a sensitizer condensation centers appear, which play the role of a nucleus of critical size on which the condensation and growth of drops of the liquid phase begin. These investigations stimulated observation of similar effects in other systems, for instance under illumination of cooled alcohol solutions of dyes. When pyrene solutions are excited in the near-ultraviolet region, their fluorescence spectrum changes [3.23]. The authors of [3.23] inferred that the change in fluorescence spectrum is caused by the formation of dimers from nonexcited and excited pyrene molecules. Formation of excited associates was discovered in a number of other investigations as well.

The authors of [3.24] observed for the first time the effect of increase in the melting temperature of the anthracene photoisomer under the effect of electron excitations. They suggested that the appearance of excitations and the formation of excited associates increases the energy of the crystal lattice and, hence, the melting temperature of the crystals. According to [3.24], the shift in the melting point of anthracene and its derivatives depends on the intensity of the "intrinsic" illumination and may be as large as tens of degrees.

Some authors reported the appearance of photocrystallization with an increase in crystal growth rate on photon irradiation. It was indicated that the growth rate of some crystals, for instance, selenium, increased on illumination. A reduction in the latent period of ice formation under the effect of ultraviolet radiation was observed [3.25]. The photocrystallization effect can be observed most clearly on illumination of solutions and melts of some aromatic hydrocarbons with photoactinic light. For a supercooled melt of a pyrene-anthracene eutectic, the photocrystallization effect is not associated with the formation of seed crystals of dianthracene. The mechanism of this phenomenon is associated, in some way or other, with the electron excitation of part of the molecules. In [3.26] it was established that illumination changes the nature of oriented crystallization of anthraquinone on the surface of single crystals of germanium. Illumination of semiconductor in the spectral region of its intrinsic photosensitivity changes both the density and orientation of the crystals. The authors named this pehnomenon photoepitaxy. The accumulated body of experimental data suggests that photoepitaxy, as well as photocrystallization in general, is based on the electron excitation mechanism, in particular, on the generation of nonequilibrium carriers by light.

Aside from the ferroelectrics described in this chapter, photosensitive phase transitions are observed in a great number of crystals and solids. We can mention here the ferromagnetic $Fe_{1-x}S$ [3.27], ferromagnetic semiconductors of the type of chalcongenides and europium oxides [3.28], some vanadium oxides, and lead oxides. According to some data, however, the transition from the orthorhombic to the tetragonal phase of PbO can be considered pyroelectric [3.29]. The photostimulated nature of this phase transition (a shift in phase transition temperature on illumination) was established in [3.29].

4. Screening of Spontaneous Polarization

In the first two chapters we discussed the role of the electron subsystem in the mechanism of ferroelectric phase transitions and the appearance of spontaneous polarization. A ferroelectric was regarded as an infinitely extended homogeneous crystal. For a real ferroelectric the boundary conditions must be determined, i.e. values of the polarization P (or displacement D) must be assigned on the surface. It is clear from energy considerations that on the surface of a ferroelectric the normal component of the spontaneous polarization must be zero; in other words, it must be screened. For ferroelectric semiconductors this screening must be internal, accomplished by electrons or holes. Nonequilibrium carriers in photoferroelectrics make an additional contribution to the screening.

The screening of the spontaneous polarization of a single-domain ferroelectric crystal determines the coordinate dependence of the bend in the bands, the displacement and the internal field. These functions, in turn, determine, on the one hand, the physical properties of the ferroelectric as an n- or p-type semiconductor, and, on the other hand, such important 'classical' properties of the ferroelectric as the switching mechanism, the distribution and value of the depolarization field, and the critical size of the ferroelectric crystal. If the screening of a single-domain crystal does not correspond to the minimum of its free energy, it is divided into domains. Thus the screening determines, among other things, the domain structure of the ferroelectric.

Therefore the nonequilibrium conductivity in photoferroelectrics, by contributing to the screening, also affects some properties of the ferroelectric, namely the domain structure, the switching mechanism, and the coercive field. Some of the phenomena in photoferroelectrics, which are associated with the screening of the spontaneous polarization by nonequilibrium carriers, are considered in Chap. 8. The present chapter deals in general terms with the screening of the spontaneous polarization.

4.1 Debye Length as Screening Length Parameter in Ferroelectrics

The Debye screening length ℓ_D is, in real ferroelectrics, always much larger than the correlation length ℓ_κ because of the high values of the dielectric constant ε. For instance, the concentration of free carriers, $n \simeq 10^{17}\ \mathrm{cm}^{-3}$, is sufficient for screening the single-domain state in $BaTiO_3$ [4.1]. At the same time, for $n \simeq 10^{17}\ \mathrm{cm}^{-3}$ and $\varepsilon \simeq 10^3$ we have $\ell_D \simeq 10^{-5}$, i.e. the Debye length is two orders of magnitude higher than the correlation length. In [4.2] it was shown rigorously that when $\ell_\kappa \ll \ell_D$, the correlation energy can be neglected as compared with the screening energy, which, in turn, makes it possible to neglect the correlation energy in the expression for the free energy. Naturally, at high concentrations, when $\ell_D < \ell_\kappa$, this approximation is unsuitable. The condition $\ell_D < \ell_\kappa$, however, corresponds to the limiting carrier concentration compatible with the existence of ferroelectricity, because then the Debye screening length is comparable with the length of the unit cell.

Thus, following [4.3], let us consider the problem of the distribution of the screening field and charge in a single-domain crystal under conditions where the Debye screening length is the length parameter. We assume the absence of surface levels. Consider the free energy[1]

$$F(D) = \frac{\alpha}{2} D^2 + \frac{\beta}{4} D^4 + \frac{\gamma}{6} D^6 + \frac{g}{4} (D_0^2 - D^2)^2 \tag{4.1}$$

together with the set of equations

$$\frac{1}{4\pi} \mathscr{E} = \frac{dF}{dD} \equiv F'(D) \quad , \tag{4.2}$$

$$\frac{dD}{dz} = 4\pi\rho \quad , \tag{4.3}$$

$$\rho = -2qn_i \sinh \frac{q\varphi}{kT} \quad , \tag{4.4}$$

---

[1] As distinct from Chap. 2, from now on the free energy will be expanded in terms of the displacement D rather than the polarization P. Since the relation $D = \quad + 4\pi P$ reduced to $D \approx 4\pi P$ for a ferroelectric, the expansions differ only by appropriate powers of 4π.

and the boundary conditions

$$D|_{z=\pm L/2} = 0 \quad , \quad D|_{z=0} = D_0 \quad , \quad \varphi|_{z=0} = 0 \quad . \tag{4.5}$$

The coefficient g is determined only by the elastic properties of the ferroelectric. The crystal is an infinitely extended homogeneous slab of thickness L, on the surface at which the boundary conditions (4.5) are met. The point Z = 0 is located at the crystal center where $D = D_0$ and no stresses are present. In (4.1-5) F is the density of free energy, D is the spontaneous displacement, $\mathscr{E}$ is the field, ρ is the charge density, φ is the potential, and n_i is the equilibrium concentration of carriers in the neutral region of the intrinsic semiconductor.

For a first-order phase transition, we have

$$\alpha = \alpha_T' \, (T - T_0) = \alpha(T_1)(1 + \frac{\Delta T}{\theta}) \quad , \tag{4.6}$$

where $\alpha(T_1) = \alpha_T'(T_1 - T_0)$, $\Delta T = T - T_1$, $\theta = T_1 - T_0$, T_1 is the phase transition temperature, T_0 is the Curie-Weiss temperature, $\alpha(T_1) > 0$, $\beta < 0$, $\gamma > 0$ and $g > 0$.

Differentiating both sides of the Poisson equation (4.3), we have

$$\mathscr{E} = - \frac{1}{4\pi \frac{d\rho}{d\varphi}} \frac{d^2 D}{dz^2} \quad . \tag{4.7}$$

Using this relation, the equation of state (4.2) takes the form

$$- \frac{1}{(4\pi)^2 \frac{d\rho}{d\varphi}} \frac{d^2 D}{dz^2} = F'(D) \tag{4.8}$$

Integrating (4.8) from D_0 to D and using the boundary conditions (4.5) gives the following expression for the screening energy:

$$- \int_0^{\varphi} \rho d\varphi = F(D) - F(D_0) \quad . \tag{4.9}$$

Substituting (4.4) into (4.9), we obtain

$$2n_i kT\left[\cosh\frac{q\varphi}{kT} - 1\right] = F(D) - F(D_0) \quad . \tag{4.10}$$

Combining (4.10), (4.3), and (4.4), we arrive at the equation describing the distribution of spontaneous displacement,

$$\frac{dD}{dz} = \pm\frac{4\pi q}{kT}\left[F(D) - F(D_0)\right]\left(1 + \frac{4n_i kT}{F(D)-F(D_0)}\right)^{1/2} \quad . \tag{4.11}$$

Let us consider two cases corresponding to the screening energy being much greater than, or much less than, the thermal energy.

In the first case $F(D) - F(D_0) \gg 4n_i kT$. Here the distribution of the volume charge density and the bend of the conduction band are given by the relations

$$\rho = \pm q\frac{F(D)-F(D_0)}{kT} \quad , \quad E_C - E_F - q\varphi = -kT\ell n\frac{F(D)-F(D_0)}{kTN_C} \quad . \tag{4.12}$$

Numerical estimates can be made in this situation, for example, for barium titanate. We find $\max[F(D) - F(D_0)] \simeq 10^6$ erg cm^{-3}, so the approximation (4.12) holds true for carrier concentrations of $n_i \ll 10^{19}$ cm^{-3}. The corresponding surface-layer carrier concentration is

$$n_s = \frac{\rho_s}{q} \simeq \frac{F(0)-F(D_0)}{kT} \simeq 10^{19}\ \mathrm{cm}^{-3} \quad .$$

In the second case $F(D) - F(D_0) \ll 4n_i kT$, and (4.11) reduces to

$$\frac{dD}{dz} = \pm\, 8\pi q\left(\frac{n_i[F(D)-F(D_0)]}{kT}\right)^{1/2} \quad . \tag{4.13}$$

This approximation corresponds to high concentrations, $n_i \geq 10^{19}$ cm^{-3}. It should, however, be remembered that this figure can represent not only free carriers in the band, but also carriers captured by shallow traps. The integral of (4.13) has the form

$$\int_t^1 \frac{dt}{\sqrt{(1-t^2)\{[2(1+\zeta)-a^2]a^2-(1+\frac{\Delta T}{\theta})+[2(1-\zeta)-a^2]a^2t^2-a^4t^4\}}} = \ell_D^{-1}(z) \ . \tag{4.14}$$

where $t = D/D_0$, $a = D_0/D(T_1)$, $\zeta = -g/\beta > 0$, and $\ell_D = (\varepsilon_c kT/16\pi q^2 n_0)^{1/2}$ is the Debye length, ε_c being the dielectric constant at the Curie point. According to (2.32), the value of the spontaneous displacement at the Curie point, $D(T_1)$, satisfies the relation

$$D(T_1) = 2\left(-\frac{\alpha_c}{\beta}\right)^{1/2} \ .$$

The expression in the square root in (4.14) is positive because the value of the displacement at the center of the crystal, D_0, satisfies the following inequality:

$$D(T_1)\left(\frac{2-(1-3\Delta T/\theta)^{1/2}}{3}\right)^{1/2} \le D_0 \le D(T_1)\left(\frac{2+(1-3\Delta T/\theta)^{1/2}}{3}\right)^{1/2} \ . \tag{4.15}$$

In (4.15) the lower limit corresponds to the maximum of the free energy (see Fig.2.3), and the upper one, to the value of (2.30) for the spontaneous displacement in an infinite crystal at an arbitrary temperature.

The solution of (4.14), obtained and analyzed in [4.3], is

$$D = D_0 \frac{\tau_1^{1/2} \mathrm{cn}y}{(\tau_1 - \mathrm{sn}^2 y)^{1/2}} \ , \tag{4.16}$$

where sny and cny are the Jacobian elliptic sine and cosine, respectively, and

$$y = a^2 \, [\tau_1 \, (1 - \tau_2)]^{1/2} \, \ell_D^{-1} \, z \ , \tag{4.17}$$

$$\tau_{1,2} = \frac{2(1-\zeta)-a^2}{2a^2} \pm \left(\frac{\zeta(\zeta-2)}{a^4} + \frac{3\zeta+1}{a^2} - \frac{\Delta T}{\theta a^4} - \frac{3}{4}\right)^{1/2} \ . \tag{4.18}$$

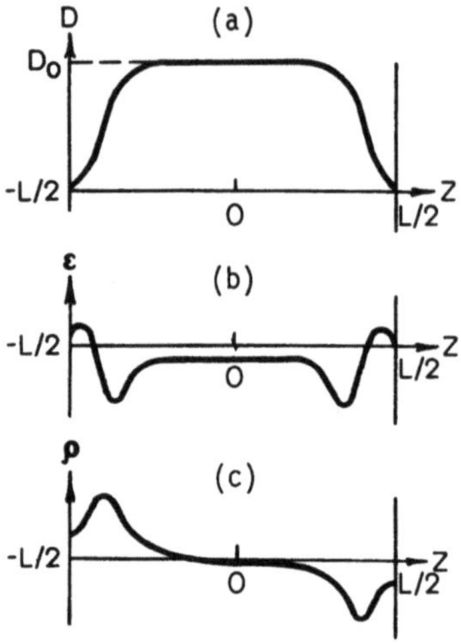

Fig. 4.1. Distribution of displacement D (a), electric field (b), and space-charge density ρ (c) in plate with single-domain polarization at $T = T_1$ [4.3]

The coordinate dependences of the spontaneous displacement, the field, and the space-charge density corresponding to (4.16) are given in Fig. 4.1. We see, first of all, that the length parameter governing the distribution of the displacement, the field, and the space charge is the Debye length ℓ_D. As can be seen from Fig. 4.1, the internal field in a single-domain ferroelectric changes sign. Within the crystal, the direction of the field is opposite to that of the spontaneous polarization. The maximum value of the depolarizing field is equal to that of the coercive field. The maximum of the screening charge density is located in the crystal volume, and the temperature dependence of its coordinate is determined by the temperature dependence of the Debye length.

The solution of (4.16) determines the critical size of the single-domain crystal with respect to the Debye length. Substituting the boundary condition (4.5) into (4.16,17) yields

$$a^2 = \left(\frac{D_0}{D_s}\right)^2 = 2\,\frac{F(q)}{[(\tau_1(1-\tau_2)]^{1/2}}\,\frac{\ell_D}{L} \quad , \tag{4.19}$$

where F is the elliptic integral, and

$$q = \left[\frac{(\tau_1-\tau_2)}{\tau_1(1-\tau_2)}\right]^{1/2}$$

is its modulus.

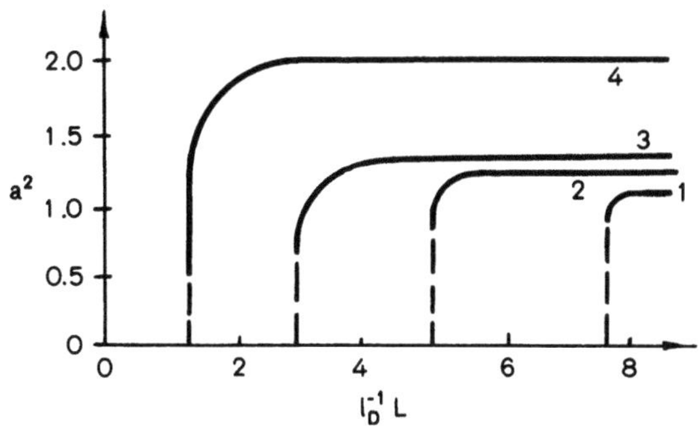

Fig. 4.2. Displacement at crystal center as a function of plate thickness for different temperatures [4.3]. 1 - $|\Delta T| = 0$; 2 - $|\Delta T| = 2.5\ \theta$; 3 - $|\Delta T| = 4\ \theta$, 4 - $|\Delta T| = 9\ \theta$

Figure 4.2 depicts the dependence of the square of the displacement at the crystal center on the relative thickness of the crystal, L/ℓ_D. The important result of this analysis consists in the fact that (as seen from Fig.4.2) there is a critical crystal thickness L_{cr} defined by the condition that at $L < L_{cr}$ no single-domain state is possible in the crystal. It is significant that L_{cr} is of the order of the Debye length and decreases with the distance from the Curie point. Then the temperature T_1^* at which the spontaneous polarization at the crystal center disappears does not, in general, coincide with the Curie point T_1. This is easy to understand if we take into consideration that the expression for the free energy of a crystal of finite size (4.1) contains the term $g(D_0^2 - D^2)^2/4$ describing the elastic energy. As would be expected, for $\zeta = -g/\beta \ll 1$ $T_1^* \simeq T_1$. For $\zeta > 1$, a finite-sized crystal may remain a ferroelectric above the Curie point since $T_1^* = T_1 + 1/3\ \theta$. It is clear from the above that this shift is directly related to the elastic stresses that result from the screening of the spontaneous displacement in a crystal of finite thickness. This effect depends only on the screening of the spontaneous displacement; in principle, size effects of a different nature may be superimposed on it, for instance the effect associated with the variation in the dispersion of the ferroelectric oscillation mode in ferroelectric films [1.7].

Similar conclusions are suggested by an analysis of the case of low carrier concentrations, $n_i \ll 10^{19}\ cm^{-3}$. Integrating (4.11) and (4.12) leads to the following expression for L_{cr}:

$$L_{cr} = 2\ell_D \quad , \quad \ell_D = \left(\frac{kT\varepsilon_c}{16\pi q^2 n^*}\right)^{1/2} \quad ,$$

$$n^* = \frac{F(0)-F(D_0)}{kT} \frac{1}{\left[\ell n \frac{F(0)-F(D_0)}{n_i kT}\right]^2} \tag{4.20}$$

From (4.20) it is clear that in this case the critical thickness L_{cr} is the same order of magnitude as the Debye length. The Debye length itself, ℓ_D, will then be determined, not by the concentration n_i in the quasineutral region of the crystal, but by the concentration n^* is the region of strong zone bending near the surface. The latter is due precisely to the screening of the spontaneous displacement at small values of n_i.

Thus, taking into account the screening of the spontaneous displacement by intrinsic carriers in a ferroelectric of finite size considerably alters the thermodynamic theory obtained in Chap. 2 for an infinitely extended crystal.

Another possible mechanism of internal screening in a ferroelectric is associated with the existence of surface layers with special dielectric properties on the ferroelectric surface. Such a layer is not a foreign film (formed, for instance, during the crystal growth), but represents a near-surface region of the crystal, in which dielectric saturation and piezoelectric constriction are observed because of the strong electric field [1.8,4.4].

The presence on the ferroelectric surface of a surface layer with dielectric properties differeing from those in the volume introduces, irrespective of the nature of the layer, specific features into the distribution of the internal field during the screening of the spontaneous displacement. In particular, the surface layer may lead to the formation of an internal depolarizing field in a ferroelectric with short-circuited electrodes. This field may be responsible for the depolarization of a single-domain ferroelectric and the influence of nonequilibrium carriers on it (the photodomain effect), short-circuit photocurrents in ferroelectrics, effects of optical damage, and many other phenomena in photoferroelectrics, which we shall discuss in the chapters to follow.

4.2 Screening and Periodic Structure of Interphase Boundaries in Ferroelectrics

In ferroelectrics which exhibit a first-order phase transition, both phases - ferroelectric and paraelectric - may coexist near the Curie point, and this leads to the formation of an interface. In the vicinity of the interface the spontaneous polarization is screened by the carriers, which results in the bending of the zones and, hence, in an 'intrinsic' field effect. The significant point here is that the screening energy must not be high compared with the elastic energy of the interface; otherwise no interface is formed and, consequently, the coexistence of the two phases in a finite temperature range becomes energetically disadvantageous.

Diffusion of carriers through the crystal may lead to volume redistribution so that a periodic or stratified structure consisting of alternating portions in the paraelectric and ferroelectric phases is formed. The regions in the paraelectric phase have a higher carrier concentration, and the regions in the ferroelectric phase, a correspondingly lower concentration due to the shift of the Curie point (see Chap.2). The periodic length of such a structure is determined by the diffusion length, since the carriers fail to diffuse over large distances during their lifetime. Thus, the equidistant system of interfaces in a ferroelectric results, on the one hand, from the diffusion-drift equilibrium in the ferroelectric and, on the other, from the interaction of the electron subsystem with the crystal lattice. This phenomenon was observed experimentally for the first time by the present author (see Chap.8) and was considered theoretically by LARKIN and KHMELNITSKY, whose work [4.5] will be discussed here.

In accordance with (2.7) we write the free energy of a ferroelectric near the Curie point as a sum of the free energy of the lattice, F_1, and the free energy of the electron subsystem, F_2, where F_1 satisfies, for instance, (4.1), and F_2 satisfies a relation similar to (2.10):

$$F_2 = E_n n_n + E_p n_p \tag{4.21}$$

where E_n and E_p are the electron and hole energies, respectively, n_n and n_p are their concentrations, and $E_g = E_n + E_p$ is the energy gap. Here, as in Chap. 2, we neglect the change in the effective masses of the carriers near the phase transition and, besides, we consider only free carriers in the bands, ignoring localized states. The equation of state (4.2) has the following form, taking (4.1,21) into account:

$$\alpha_T'(T - T_0)P - |c|P^3 + fP^5 - \kappa \frac{d^2P}{dz^2} + n_n \frac{dE_n}{dP} + n_p \frac{dE_p}{dP} - \mathscr{E} = 0 \quad . \tag{4.22}$$

The coefficients in (4.72) differ from those in (4.1) by the factor 4π raised to the appropriate power.

The electric field $\mathscr{E}$ satisfies Poisson's equation

$$\mathrm{div}(\mathscr{E} + 4\pi P) = 4\pi q(n_n - n_p) \quad . \tag{4.23}$$

Let us consider the temperature range near the phase transition,

$$\Delta T \simeq T_1 - T_0 \simeq 3\,\frac{c^2}{4\alpha_T' f} \quad ,$$

in which the two phases can coexist. This possibility depends on the electron concentration, which also determines the position and concentration of the interfaces. To see this, supplement (4.22,23) with the continuity equation

$$\frac{d\rho}{dt} + \frac{dj}{dz} = 0 \tag{4.24}$$

where ρ is the charge density, and j is the current density. Substituting the expression for the current and the charge density for the electrons and holes, respectively, into (4.24), we arrive at the following equations:

$$\begin{aligned} &\frac{d}{dz}\left[D_n\left\{\frac{dn_n}{dz} + n_n\left(\frac{dE_n}{dz} + qE\right)\Big/kT\right\}\right] + G - R = 0 \\ &\frac{d}{dz}\left[D_p\left\{\frac{dn_p}{dz} + n_p\left(\frac{dE_p}{dz} - qE\right)\Big/kT\right\}\right] + G - R = 0 \quad , \end{aligned} \tag{4.25}$$

where D_n and D_p are, resepctively, the electron and hole diffusion coefficients, G and R are, respectively, the generation and recombination rates of nonequilibrium electron-hole pairs.

Let us look for periodic solutions of the set of equations (4.22, 23, 25), assuming that the period exceeds the correlation length $\ell_\kappa = [\kappa/\alpha_T'(T_1 - T_0)]^{1/2}$ and the Debye length ℓ_D; we also assume that $L_D \gg \ell_\kappa, \ell_D$, where $L_D = (D\tau)^{1/2}$ is the diffusion length (τ is the lifetime). Near the interfaces we can neglect the current across the boundary and the last two terms in (4.25). This leads to the following values of the concentration:

$$n_n \simeq \exp\left[-\frac{E_n+q\varphi}{kT}\right] \quad , \quad n_p \simeq \exp\left[-\frac{E_p+q\varphi}{kT}\right] \quad , \tag{4.26}$$

where φ is the potential. For distances from the boundary exceeding ℓ_κ and ℓ_D ($q\varphi \ll E_n, E_p$) we obtain the electroneutrality condition $n_n = n_p = n/2$ from (4.23) and (4.26). For the ratio of the carrier concentration in the ferroelectric and paraelectric phase (on both sides of the boundary) we have

$$n^s/n^p = \exp(-\Delta E_g/2kT) \quad , \tag{4.27}$$

where $\Delta E_g = E_g^s - E_g^p$ is the jump of the energy gap on a first-order phase transition. Away from the interface we can neglect the derivatoves dE_n/dz and dE_p/dz, and thus the system (4.25) is reduced to the following form with an allowance for $n_n = n_p$:

$$\frac{d^2n}{dz^2} + \frac{1}{D}(G - R) = 0 \tag{4.28}$$

where the bipolar diffusion coefficient $D = 2(D_n + D_p)^{-1}$. If we substitute into (4.28) the rate of recombination $R = n/\tau$ and the rate of generation $G = n_0/\tau$, where n_0 is the generated carrier concentration proportional to the light intensity and the quantum yield, (4.28) takes for form

$$\frac{d^2n}{dz^2} + \frac{1}{L_D^2}(n_0 - n) = 0 \quad . \tag{4.29}$$

Here, L_D is the diffusion length, which depends, via D and the lifetime τ, on the phase in which the carriers are excited. The solution of (4.29) must be in line with the boundary conditions (4.27) and

$$D^s \frac{dn^s}{dz} = D^p \frac{dn^p}{dz} \quad . \tag{4.30}$$

The last condition represents the equality of the diffusion flows across the interface.

The boundary equilibrium condition in the presence of electrons can be obtained form (4.22). Multiplying (4.22) by dP/dz and integrating with respect to z in the range of values $\ell_\kappa, \ell_D < z < L_D$, we obtain, using (4.23),

$$F^s(P) + kTn^s + \frac{(\mathscr{E}^s)^2}{8\pi} = F^p(P) + kTn^p + \frac{(\mathscr{E}^p)^2}{8\pi} \quad , \tag{4.31}$$

where $\mathscr{E}^s$ and $\mathscr{E}^p$ is the electric field in the ferroelectric and paraelectric phase, respectively. Thus, the boundary equilibrium condition represents the equality of the pressure of the two phases on it. Neglecting the electric field pressure in (4.31) and expressing $F^s - F^p$ in terms of the heat of transition ΔQ, we have, using (4.27)

$$\frac{\Delta Q}{T_1^2}(T_1 - T) = n^s\left[\exp\left(\frac{\Delta Eg}{2kT}\right) - 1\right] \quad . \tag{4.32}$$

The concentration n^s appearing in (4.32) and equal to n_0 to an order of magnitude is determined from the diffusion equation (4.29). The relation (4.32) can be regarded as the dependence of the temperature range $T_1 - T$, in which interfaces exist, on the carrier concentration $n^s \simeq n_0$. The temperature range in which interfaces exist widens with an increase in the concentration of nonequilibrium carriers. Below some critical carrier concentration n_{cr}, on the contrary, no interfaces whatever must be observed on the finite temperature range. In order to determine the temperature range in which the interface exists from (4.32), we find an expression for n^s that appears in this equation. Solving (4.29) for the above boundary conditions yields

$$n(z) = n_0^s - \frac{n_0^s - n_0^p \exp(-\Delta E_g/2kT)}{1+\sqrt{D^s\tau^p/D^p\tau^s}\exp(-\Delta E_g/2kT)} \exp\left(\frac{z}{L_D^s}\right) \qquad \text{for } z < 0 \quad ,$$

$$n(z) = n_0^p + \frac{n_0^s \exp\left(\frac{\Delta E_g}{2kT}\right) - n_0^p}{1+\sqrt{D^p\tau^s/D^s\tau^p}\exp(\Delta E_g/2kT)} \exp\left(-\frac{z}{L_D^p}\right) \qquad \text{for } z > 0 \quad . \tag{4.33}$$

Here, $z < 0$ corresponds to the ferroelectric phase, $z > 0$ to the paraelectric phase, and $z = 0$ to the boundary; $n_0^s = n_{z=-\infty}$, $n_0^p = n_{z=+\infty}$. Using the solution of (4.33) and putting $D^s \simeq D^p$, we get the following expression for the carrier concentration near the boundary, n_s:

$$n_s = n_0^s \frac{\alpha}{1+\alpha}\left(1 + \sqrt{\frac{\tau^p}{\tau^s}}\right) \quad , \tag{4.34}$$

where $\alpha = (\tau^p/\tau^s)^{1/2} \exp(-\Delta E_g/2kT)$. This expression takes into account that $n_0^p = (\tau^p/\tau^s)n_0^s$. From now on we shall omit the superscript in n_0^s. Substituting (4.34) into (4.32), we determine the phase transition temperature

$$T_1' = T_1 - \frac{T_1^2}{\Delta Q} n_0 \frac{\alpha}{1+\alpha}\left(1 + \sqrt{\frac{\tau^p}{\tau^s}}\right)\left[\exp\left(\frac{\Delta E_g}{2kT} - 1\right)\right] \quad . \tag{4.35}$$

Here, $T_1 - T_1'$ represents the shift of the phase transition temperature due to the contribution of the electron subsystem. From (4.35) it is seen that this shift is directly proportional to the carrier concentration n_0. Thus, for $\Delta E_g \ll kT$ and $\tau^s \simeq \tau^p$, (4.35) coincides with (2.36). In the n_0 - T diagram (Fig.4.3) the phase transition is shown as a dashed line.

From (4.35) it follows that with a given carrier concentration n_0, a single interface can be at equilibrium at one definite temperature T_1'. For a system of interfaces there must be a finite temperature range T_1 - T. This is easy to understand from (4.29) and (4.34). The concentration n^s depends not only on the light intensity, but also on the distance between the boundaries. A group of boundaries must form an equidistant structure. This follows from (4.29) and the boundary condition (4.27), because only the periodic solution of (4.29) gives the same value of n^s at each boundary. The boundary equilibrium condition (4.31) determines the ratio between the dimensions x and y of the ferroelectric and paraelectric regions of this structure. From the

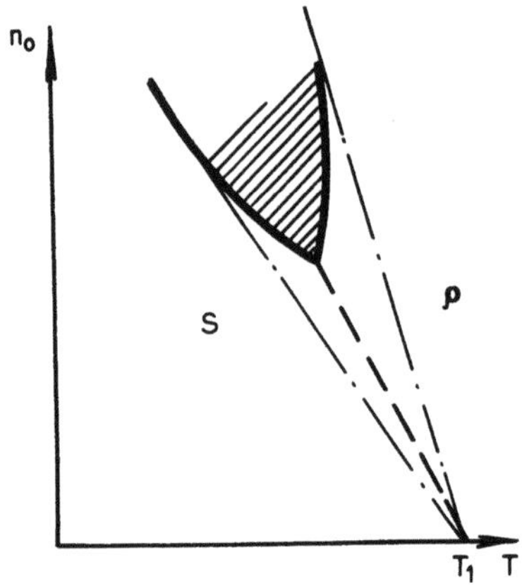

Fig. 4.3. n_0 - T state diagram of coexisting phases [4.5]

periodic solution of (4.29) we determine n^s and, substituting this value into the equilibrium condition (4.32), we get

$$\tanh \frac{x}{2L_D^s} \coth \frac{y}{2L_D^p} = \frac{1}{\alpha} \frac{t-\alpha(\tau^p/\tau^s)^{1/2}}{1-t} \quad , \tag{4.36}$$

where

$$t \equiv \frac{(T_1-T)\Delta Q}{T_1^2 n_0\left[\exp\left(\frac{\Delta E_g}{2kT}\right)-1\right]} \quad . \tag{4.37}$$

From (4.36) it is seen that the periodic structure of interfaces can exist in the temperature range

$$\frac{T_1^2 n_0\left[\exp\left(\frac{\Delta E_g}{2kT}\right)-1\right]}{\Delta Q} > T_1 - T > \frac{T_1^2 n_0\left[\exp\left(\frac{\Delta E_g}{2kT}\right)-1\right]}{\Delta Q} \alpha \sqrt{\frac{\tau^p}{\tau^s}} \quad . \tag{4.38}$$

In the phase of diagram, Fig. 4.3, this region is bounded by dot-and-dash lines.

The critical concentration at the transition temperature below which the stratified structure is metastable at all temperatures can be found as

$$n_k = \frac{\sigma(1-\alpha^2)}{L_D^s T_1 \left[\exp\left(\frac{\Delta E_g}{2kT}\right)-1\right] \alpha(1-\alpha\sqrt{\tau^p/\tau^s})\,\ell n[2/(1+\alpha)]} , \qquad (4.39)$$

where σ is the surface tension of the boundary. The period structure does not exist at $n < n_k$ inside, the period decreases with increasing concentration n.

Actually, the time taken to establish equilibrium may be large (much larger than the time to diffusion-drift equilibrium). This contributes to the hysteresis on a first-order phase transition.

5. Photoferroelectric Phenomena and Photostimulated Phase Transitions

Photoferroelectric phenomena encompass a large class of effects with a wide variety of mechanisms. Nevertheless, they can be divided into two main groups. The first group includes phenomena associated with the excitation of nonequilibrium carriers and the optical recharging of levels in the neutral volume of the ferroelectric with no internal field present [3.6]. This group of phenomena is described by the thermodynamic relations of Sect. 2.3 and by the mechanisms considered in Chap. 3, in particular, the pseudo-Jahn-Teller effect. As shown in Chap. 4, however, a real ferroelectric always has an internal field, whose value and distribution depend on the screening of the spontaneous polarization in the volume and on the surface of the crystal. Excitation of nonequilibrium carriers, changing the screening conditions, simultaneously changes the value and distribution of the internal field in the ferroelectric. This may result in a shift of the Curie point and a change in the whole set of properties of the ferroelectric, including its domain structure. This second group of phenomena due to the influence of the spontaneous polarization will be conventionally called screening phenomena. Thus, the ferroelectric phenomena that are actually observed may be associated with two completely different mechanisms, whose experimental differentiation usually involves difficulties. In most cases the influence of intrinsic illumination on the physical properties of a ferroelectric is due to the effects of both mechanisms.

In this chapter we shall mainly discuss the first group of photoferroelectric phenomena, although many of them (as we shall see later) represent a complex superposition of phenomena from both groups. Chapter 8 is concerned with photoferroelectric phenomena of the second group, namely screening phenomena.

5.1 Thermodynamics of Photoferroelectric Phenomena

Near a first-order phase transition, the effect of the change in hysteresis ΔT_{hN} is superposed on the photostimulated shift of the phase transition temperature ΔT_{1N} (the photohysteresis effect). Therefore, for a first-order phase transition ΔT_{1N} is not the same as the quantity ΔT_N determined by the expression (2.36). Besides, the values of ΔT_{1N} and ΔT_{hN} may be close to the temperature range near the phase transition in which the polarization fluctuation cannot be neglected and, hence, the expansion (2.14) cannot be used. All this must be modified to take into account the effects described in Sect. 2.3.

In the case of first-order phase transitions the coefficient α in the expansion of the thermodynamic potential does not reduce to zero at the transition point T_1. From the derivation of the relation (2.36) it is clear that for first-order phase transitions, (2.36) describes the photostimulated shift of the Curie-Weiss temperature T_0, which is determined from the condition $\alpha(T = T_0) = 0$ and appears in the Curie-Weiss law. The Curie-Weiss temperature and its shift on illumination can, generally speaking, be determined experimentally from the temperature curve $1/\varepsilon$ in the paraelectric region. In practice, however, such measurements must be highly accurate. Indeed, from the Curie-Weiss law (2.28) it follows that the change in ε due to a change in Curie-Weiss temperature satisfies the relation

$$\frac{\Delta\varepsilon}{\varepsilon} = \frac{\Delta T_0}{T-T_0} \quad .$$

If we assume that the difference $T_1 - T_0 \simeq 10^\circ C$, as, for example, in SbSI and $BaTiO_3$, and $\Delta T_0 \simeq 1^\circ C$, then it is clear that even in the immediate vicinity of the phase transition $\Delta\varepsilon/\varepsilon \simeq 0.1$, and that it decreases with rising temperature.

For a first-order phase transition the shift of the transition point on a change in the concentration of nonequilibrium carriers can be obtained from (2.35):

$$\Delta T_1 \simeq -\frac{C}{2\pi} aN + \left(\frac{2bN}{\beta} - \frac{cN}{\gamma}\right)\frac{1}{8\pi}\,\frac{\beta^2 C}{\gamma} \quad . \tag{5.1}$$

Thus, if the shift ΔT_0 is determined only by the coefficient a, then the shift of the phase transition temperature is additionally determined by the coefficients b and c. In Sect. 2.3 we showed that for ferroelectric phase transitions a > 0 and hence $\Delta T_0 < 0$. At the same time thermodynamics does not predict the sign of the coefficients b and c, and both situations are possible, namely $\Delta T_1 < 0$ and $\Delta T_1 > 0$. Thus, for first-order phase transitions, when estimating the photostimulated shift of the phase transition temperature one should also bear in mind the photohysteresis effect.

Estimation of the temperature hysteresis of the upper phase transition in $BaTiO_3$ from (2.35) using experimental values of β, γ and the Curie-Weiss constant C leads to a figure eight times greater the observed value ΔT_h [5.1]. A similar calculation for SbSI gives $\Delta T_h = 6.5^\circ C$, whereas the experimentally observed value is less than $1^\circ C$. The inadequacy of conventional thermodynamic theory in considering the temperature hysteresis of the phase transition stems from neglect of fluctuations. The problem of polarization fluctuations near a ferroelectric phase transition was first considered by GINZBURG [2.10]. According to this paper, the thermodynamic theory of ferroelectrics based on the expansion (2.14) is valid only for a definite region near the phase transition, with the exception of the temperature range near this point, namely:

$$\frac{\Delta T}{T_1} \gg \frac{k^2 T_1 \beta^2}{32\pi^2 \kappa^3 \alpha'_T} \quad . \tag{5.2}$$

Here, ΔT is the value of the range where the expansion (2.14) fails to apply, and κ is the correlation parameter.

Estimates show that at a temperature of the order of one degree from the phase transition point the expansion (2.14) may become incorrect.

Of still more importance is the inclusion of the heterophase fluctuations near a first-order phase transition temperature. It is well known that in $BaTiO_3$, nuclei of the new phase appear and disappear irregularly as we approach the point of the upper phase transition [5.1]. If the new phase is stable at a given temperature, then sufficiently large nuclei of the new phase are stable and continue to grow with time. The minimum ('critical') size of a stable nucleus is determined by the following condition. In the phase transition the reduction in free energy due to the formation of a certain volume of the new phase is offset by an increase in free energy due to the appearance of the interface. Thus, there is a minimum size which the nucleus of the new phase must have for its growth to be thermodynamically advantageous. The formation of such nuclei actually determines the temperature hysteresis of the phase transition. It is clear that the temperature hysteresis ΔT_h^f taking into account the fluctuation is less that the value of ΔT_h found from (2.35). If $\Delta T_h^f \ll \Delta T_h$, the photostimulated change ΔT_h, i.e. the decrease in the region of existence of metastable states caused, for instance, by illumination, may not be observed at all and the conclusions of Sect. 2.3 do not apply to this case.

The influence of nonequilibrium carriers on the temperature hysteresis of the phase transition, ΔT_h^f, may be due to the following mechanism. The probability of formation of a spherical nucleus of critical size at a fixed value of superheating (or supercooling) depends on the value of the surface energy of the nucleus, W, as $\exp(-W^3)$ [2.8]. In the case of a ferroelectric phase transition, when one phase is polar, W is the sum of the elastic energy $\sigma = W_1$ due to the distortion of the crystal lattice at the interface, and the screening energy of spontaneous polarization, W_2:

$$W = W_1 + W_2 \tag{5.3}$$

Here, the spontaneous polarization is screened near the interface by free (or trapped) carriers. The energy of screening W_2 of the spontaneous polarization P by carriers with a density of ρ is determined by integrating the free energy of the crystal, F, up to the Debye length ℓ_D:

$$F = \frac{1}{2}\alpha P^2 + \frac{1}{4}\beta P^4 + \frac{1}{6}\gamma P^6 \quad , \quad W_2 = \int_0^{\ell_D} F dz$$

using Poisson's equation

$$\frac{dP}{dz} = \rho$$

and the boundary conditions

$$P_{z=0} = 0 \quad , \quad P_{z=\ell_D} = P_0 \quad , \quad F(P_0) = 0 \quad , \quad \frac{dF}{dP}(P_0) = 0 \quad ,$$

where α, β, and γ are known coefficients in the expansion (2.14). As in Chap. 4, we neglect the correlation energy in F. The result is

$$W_2 = 4 \times 10^{-2} \frac{\alpha}{\rho} P_0^3 \text{ erg cm}^{-2} \quad . \tag{5.4}$$

Here, α is the inverse dielectric constant (1.26), $\rho = (N + n)q \simeq Nq$ is the screening charge density, and n and N are, respectively, the concentration of free carriers and of carriers localized in the traps (in ferroelectrics $N \gg n$). From (5.4) it is clear that the increase in the concentration of nonequilibrium carriers when a ferroelectric is illuminated in the region of its intrinsic photosensitivity reduces W_2 and, hence, the temperature hysteresis. At high light intensities and a correspondingly high filling of the traps, the screening energy W_2 may become less than the elastic energy W_1, and, according to (5.3), the dependence of the value of the temperature hysteresis on the light intensity must reach saturation. This saturation, in principle, may also be due to the complete filling of the traps. On the other hand, illumination of a ferroelectric in the impurity spectral region, which empties the traps due to the transition of the captured carriers to the band, increases W_2 and thus brings ΔT_h to the equilibrium value. The photohysteresis effect thus belongs to the group of screening phenomena.

5.2 Photostimulated Shift of Phase Transition Temperature and Photohysteresis Effect

The influence of the electrons on the Curie-Weiss temperature and the temperature hysteresis of the phase transition discussed in Sect. 2.3 are best observed experimentally by exciting nonequilibrium carriers. A photostimulated shift of a phase transition was first observed for crystals of SbSI [5.2,3], and later in an extensive series of investigations [5.4-6]. Figure 5.1 shows the dependences of the capacitance on temperature for $BaTiO_3$ crystals with a cobalt impurity, which indicate that illumination of the crystal in the intrinsic absorption region leads to a 5°C shift of the Curie point toward the lower temperatures [5.6]. Unfortunately, Fig. 5.1 does not contain any data on the temperature hysteresis and its change on illumination. The problem of differentiating photoferroelectric phenomena of different kinds in $BaTiO_3$ will be discussed later. Here we consider only photoferroelectric phenomena in SbSI and in solid solutions based on type-$A^{V}B^{VI}C^{VII}$ compounds.

In conformity with the conclusions of the thermodynamic theory, illumination of SbSI crystals results in a shift of the Curie point toward lower temperatures, and the maximum value of the shift is 1-1.5°C. The shift of the Curie temperature on illumination, ΔT_N, depends on the sum n + N, where n and N are, respectively, the concentrations of free carriers and carriers in the traps. Since for SbSI, as well as for other high-resistance photoconductors, the trapping coefficient $\theta = n/N \ll 1$, the contribution of the free carriers to the photosensitive shift of the Curie point can be neglected, and hence ΔT_N depends solely on N [by (2.36)], where N denotes the concentration of localized electrons. The measurements of ΔT_N for one of the SbSI crystals [5.2] lea to a value of $N \simeq 3 \times 10^{17}$ cm^{-3}, which is, in principle, authentic enough. An attempt to establish the relationship between the value of the shift of the Curie point on illumination, ΔT_N, and the filling of the traps was made in [5.2]. For this purpose two groups of independent measurements were carried out. Firstly, the dependence of the Curie-point shift ΔT_N on the light intensity was measured. Secondly, the method of thermostimulated currents was used to determine the activation energy of the traps and their concentration. The results of these independent measurements agree fairly well.

Single needle-shaped crystals of SbSI grown from the gas phase were also investigated (the needle axis coincided with the direction of spontaneous polarization). The Curie point of these crystals was near 23°C, and the main ferroelectric parameters were in good agreement with the published data [1.14]. The intrinsic-absorption edge and the photoconductivity maximum of

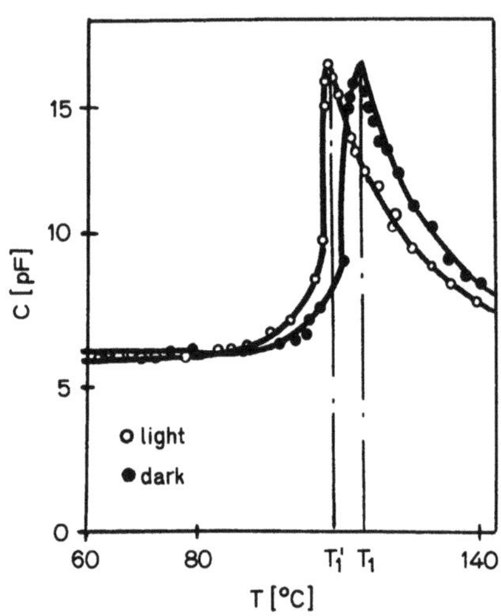

Fig. 5.1. Influence of intrinsic illumination on phase transition temperature in $BaTiO_3$ measured from temperature dependence of capacitance [5.6]

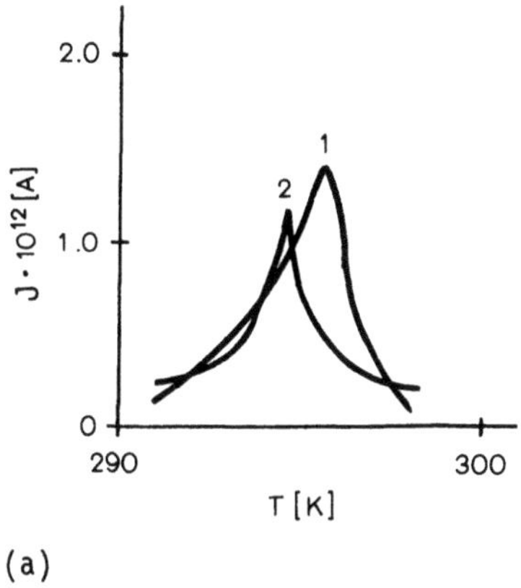

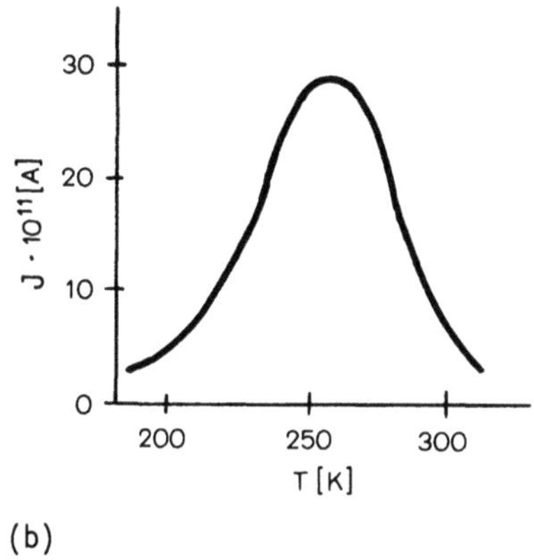

Fig. 5.2. Temperature dependence of pyroelectric and thermostimulated currents in SbSI. (a) Maxima of pyroelectric current in the dark (1) and under illumination (2), $\Delta T_1 \simeq 1$ K; (b) Thermostimulated current

the SbSI crystals are in the range of 620-660 nm. In accordance with (2.47), in a phase transition the energy gap of SbSI experiences a jump, $\Delta E_g \simeq 10^{-2}$ eV [1.14].

As distinct from [5.2,3], in [5.4] the shift of the Curie point on illumination, ΔT_N, was determined by measuring the pyroelectric current. Figure 5.2a presents two plots of the pyroelectric current taken in darkness (curve 1) and on illumination of the crystal in the region of the photosensitivity maximum (curve 2), which illustrate the illumination-induced shift of the pyrocurrent maximum towards the lower temperatures.

The measurements of the thermostimulated current in SbSI were conducted by the usual procedure [5.7]. Figure 5.2b shows the thermostimulated current plotted with due regard for the contribution from the pyrocurrent and showing one maximum at -20°C. The activation energy of the levels was determined from the maximum of the thermostimulated current with the aid of the Bube equation, and their concentration was estimated by integrating the thermostimulated current and determining the photoelectric yield [5.7]. The results of these measurements were independent of the intensity and time of illumination of the crystal, which pointed to a complete filling of the levels during the measurement of the thermostimulated current. Similar data on measurements of the thermostimulated conductivity for SbSI were obtained independently.

The spectral distribution of the photostimulated shift of the Curie point has been investigated. Figure 5.3 shows the dependence of the shift of the Curie point, ΔT_N, per unit incident energy on the wavelength of the indicent light (curve 2). For comparison, the same figure shows the spectral

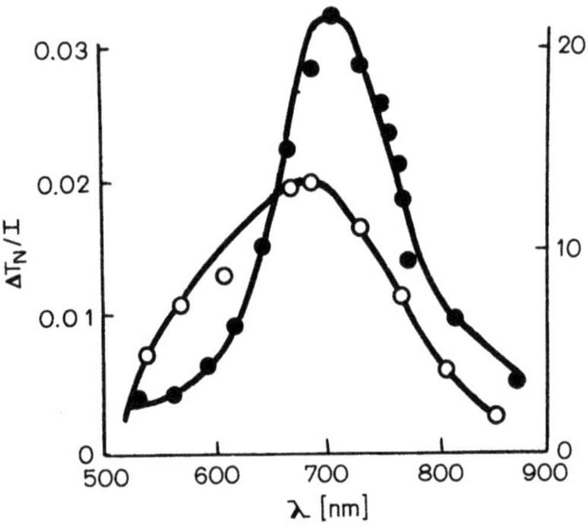

Fig. 5.3. Spectral distribution of photoconductivity (1) and of effect of photosensitive shift of Curie point (2) for SbSI (in arbitrary units)

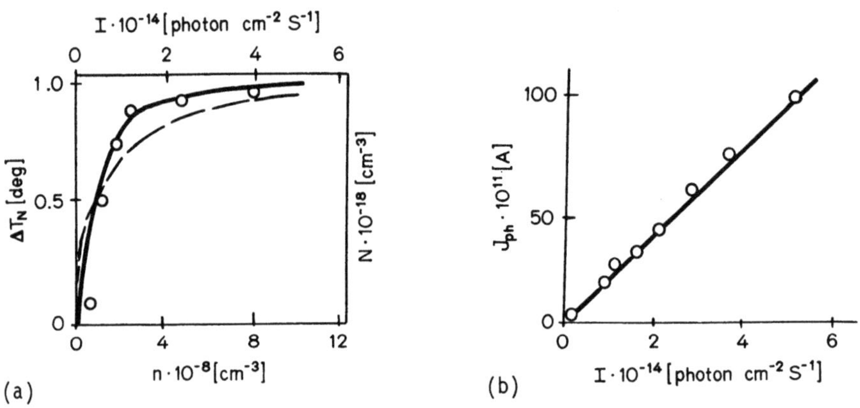

Fig. 5.4. Dependence of Curie point shift ΔT_N and photocurrent J_{ph} on intensity of light incident on crystal ($\lambda \simeq 630$ nm). (a) Dependence of ΔT_N on I. Continuous curve - experimental values of ΔT_N; dashed curve - theoretical values calculated by Eq. (4.6); values of N calculated by Eq. (4.7) are laid off on right-hand vertical axis; (b) dependence of J_{ph} on I; crystal temperature 16°C

distribution of the photoconductivity (curve 1). In conformity with theory, the maxima of both curves are located near 650 nm, i.e. the shift of the Curie point on illumination of SbSI occurs in the region of the intrinsic photosensitivity of the crystal.

Figure 5.4a exhibits the dependence of the shift of the Curie point, ΔT_N (in degrees) on the intensity, I, of the monochromatic light incident on the crystal ($\lambda \simeq 630$ nm). Fig. 5.4b portrays the current-intensity characteristic of the photocurrent for the same range of light intensities. A comparison of the curves in Fig. 5.4 shows that a crystal with a linear current-intensity characteristic shows saturation in the ΔT_N vs I curve. The comparison of the results of measurements of the thermostimulated current and the shift of the Curie point on illumination, which is given below, indicates that the saturation of the curve in Fig. 5.4a is due to the complete filling of the traps in SbSI. Restricting ourselves to the quadratic term with respect to polarization in (2.46), we have from (2.36)

$$\Delta T_N = \frac{\Delta E_g C}{\pi P_0^2} N \quad , \tag{5.5}$$

where P_0 is the spontaneous polarization, C is the Curie-Weiss constant, ΔE_g is the energy-gap jump in a first-order phase transition, and N is the carrier concentration in the traps. Substituting into (5.5) the experimentally measured value of the shift, $\Delta T_N \simeq 1^\circ C$, corresponding to the saturation of the curve in Fig. 5.4a, and also the independently measured values of $P_0 \simeq 10\ \mu C\ cm^{-2}$, $C \simeq 1.1 \times 10^5$ K, $\Delta E_g \simeq 1.0 \times 10^{-2}$ eV, we obtain $N \simeq 1.2 \times 10^{18}\ cm^{-3}$.

The concentration N can also be estimated from data on the thermostimulated current. According to [5.8], N = Q/q VG, where Q is the total charge, determined by integrating the thermostimulated current (Fig. 5.2b), V is the volume of the crystal, q is the elementary charge, and $G = J_{ph}/I\ Vq$ is the photoelectric yield, which depends on the ratio of the photocurrent to the light intensity, J_{ph}/I, i.e. on the slope of the current-intensity characteristic Fig. 5.4b. Substituting $Q \simeq 4 \times 10^{-8}$ C and $G \simeq 4.6 \times 10^{-4}$, we get $N \simeq 1.1 \times 10^{18}\ cm^{-3}$. Thus, the values of the electron concentration in the traps determined from the shift of the Curie point and from the data on the thermostimulated current are in good agreement. From this, in turn, we conclude that the value of the limiting shift ΔT_N corresponding to the saturation of the curve in Fig. 5.4a actually does correspond to complete filling of the levels, and the intermediate values of ΔT_N, to partial filling of the levels.

The activation energy of the levels can be determined by the Bube equation from the maximum of the thermostimulated current,

$$u = kT_m \ln \frac{N_c q \mu V S}{J_m \ell} \quad ,$$

where T_m and J_m are, respectively, the temperature and the thermostimulated current corresponding to the maximum of the curve in Fig. 5.2b; S and ℓ, respectively, are the cross-sectional area and the length of the crystal in the direction of the applied current, V is the voltage applied to the crystal, $\mu \simeq 50\ cm^2\ V^{-1}\ s^{-1}$ is the mobility of the electrons in SbSI [5.9], and $N_c \simeq 10^{19}\ cm^{-3}$ is the density of states. Substituting the values of the parameters corresponding to Fig. 5.2b into the equation, we find $u \simeq 0.6$ eV.

The dependence on the light intensity I of the concentration of electrons in the traps, N, can be obtained, as usual, from the stationary solution of the kinetic equation (α trapping [3.12])

$$N = \frac{n}{N_{cM}+n} M \tag{5.6}$$

where M is the concentration of the levels, n in the concentration of free electrons, $N_{cM} = N_c \exp(-u/kT)$. The dependence of N on the light intensity is expressed in (5.6) through n. At $n \gg N_{cM}$ we have saturation $N = M$, i.e. complete filling of the levels. Substituting the experimental values of n and N_{cM} into (5.6), we can obtain the dependence of N on I and compare it with the curve of Fig. 5.4a. Since the values of N in Fig. 5.4a have been obtained from the measurements of the shift ΔT_N, using (5.5), this comparison serves as an additional check on the theory. Substituting into (5.6) the values $N_{cM} \simeq 6 \times 10^7\ cm^{-3}$, $M \simeq 10^{18}\ cm^{-3}$ determined from data on the thermostimulated current, and also the values of the concentration of free carriers determined from the photocurrent for $\mu \simeq 50\ cm^2\ V^{-1}\ s^{-1}$ (Fig.5.4b), we find the dependence of N on I. This dependence is represented by the dashed curve in Fig. 5.4a and, as can be seen from the figure, the curves are close to each other.

The results indicate good agreement between the data of the measurement of the photosensitive shift of the Curie temperature, the photoconductivity and the thermostimulated conductivity of SbSI, which in turn serves as experimental proof of the validity of (2.36) and the other conclusions of the theory. Note that this agreement could be worse, in principle. This is associated not only with the necessity of rigorous determination of the carrier lifetime when estimating the concentration of the levels from the data on the thermostimulated current. Some discrepancy may be due to the uncertainty in the estimate of the parameter P_0 and of the energy-gap jump ΔE_g. The uncertainty in the estimate of the values of spontaneous polarization P_0 becomes substantial in the case of not-very sharp phase transitions. In deriving (2.36) the activation energy of the traps was neglected as compared with the energy gap. Inclusion of the energy of the traps both for the electrons and for the holes leads to a correction in the estimate of ΔE_g whose sign and value depend on the shift of the traps on phase transition. The estimate shows that this correction may be substantial. Comparison with (2.36) is justified because the level with $u \simeq 0.6$ eV is in thermal equilibrium with the band at the temperature of the phase transition in SbSI (Sect.3.4).

Figure 5.5 illustrates the coincidence of the maximum of the spectral distribution of photoconductivity and the minimum of the value of the temperature hysteresis. Measurements carried out at equal energies confirm that the photohysteresis effect is associated with the increase in crystal photoconductivity. Figure 5.6 exhibits the dependence of the temperature hysteresis on the photoconductivity obtained on a change in illumination intensity in the region of the intrinsic absorption edge. This dependence shows saturation.

To compare the experimental results with theory we determined the spontaneous polarization P_0 and, from the thermostimulated conductivity, the concentration of electrons in the traps, N. For the crystal investigated in Figs. 5.5,6, we obtained values of $P_0 \simeq 15\ \mu C\ cm^{-2}$, $M \simeq 10^{18}\ cm^{-3}$, and $u \simeq 0.55$ eV,

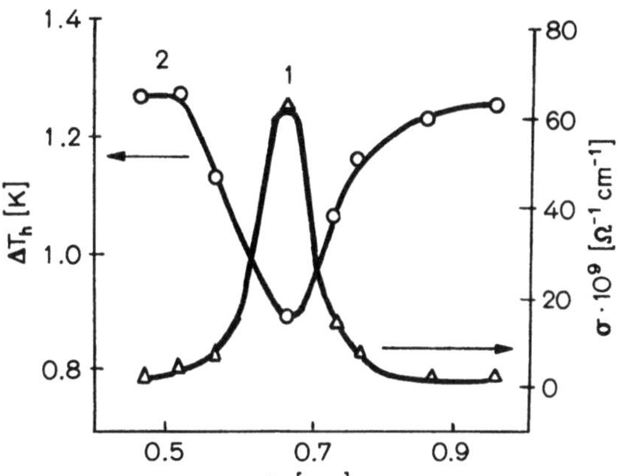

Fig. 5.5. Spectral distribution of photoconductivity (1) and of photohysteresis effect (2) for SbSI

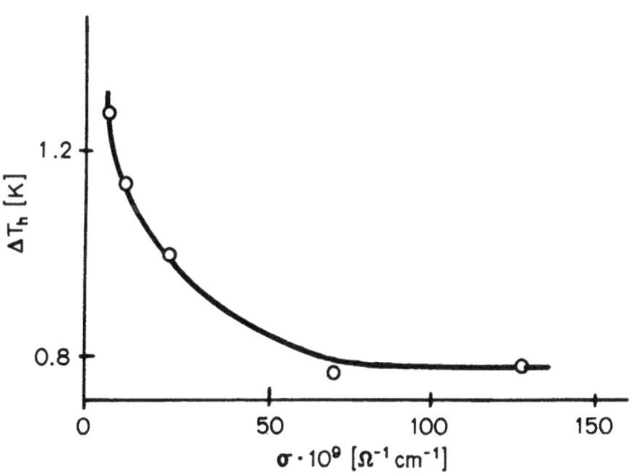

Fig. 5.6. Dependence of temperature hysteresis on photoconductivity for SbSI

where M is the concentration of the traps, and u is their activation energy. The concentration of electrons in traps, N, was determined by (5.6). The concentration of free electrons, n, which appears in (5.6), was determined from the stationary photoconductivity. In the saturation portion of the photohysteresis effect $\sigma \simeq 10^{-7}\ \Omega^{-1}\ cm^{-1}$ (see Fig.5.6). Assuming, as before, the electron mobility in SbSI to be $\mu \simeq 50\ cm^2\ V^{-1}\ s^{-1}$, we have $n \simeq 10^{10}\ cm^{-3}$, and, from (5.6), $N \simeq 10^{18}\ cm^{-3}$. Thus, in the saturation portion of the photohysteresis effect $N \simeq M$, and hence complete filling of the traps is one of the possible causes of saturation. An estimate of the screening energy using (5.4) in the saturation portion under the same conditions yields $W_2 \simeq 10$ erg cm^{-2}. It can be seen that it agrees fairly well with the other condition of saturation of the photohysteresis effect $W_2 \ll W_1$, since for SbSI the elastic energy of the nucleus boundary $W_1 \simeq 100$ erg cm^{-2} estimated from independent experiments ([5.10]; see also Chap.8). Note that this value of W_1 coincides to an order of magnitude with the value of W_1 obtained for perovskite ferroelectric semiconductors [5.11]. Thus, the data obtained for SbSI do not enable us to differentiate between the two possible mechanisms of saturation of the photohysteresis effect. Because of the lack of reliable values of the coefficient b in (2.46), it is also impossible to compare the experimental values of ΔT_{hN} for SbSI with (2.39).

After the light is switched off the temperature hysteresis increases slowly to the dark equilibrium value- it would be natural to associate this with thermal emptying of the traps. In SbSI one also observes the phenomenon of optical quenching of the photohysteresis effect, which was mentioned in Sect. 5.1 and which consists in emptying the traps by longwave radiation. To observe this, the temperature hysteresis of the phase transition was measured under illumination of the crystal in the intrinsic photosensitivity region and simultaneously in the infrared range. Quenching of the photohysteresis effect was demonstrated on additional illumination with a wavelength of 2.1 μm. Figure 5.7 illustrates the dependence of ΔT_h on the wavelength of the infrared illumination. It is seen that quenching does take place, and the quenching maximum corresponds to a wavelength of $\lambda \simeq 2.1$ μm, where the value of the temperature hysteresis almost reaches the dark equilibrium value. The activation energy of the traps, determined from the spectral distribution of the quenching effect, is 0.6 eV, which agrees fairly well with the data on the thermostimulated conductivity. Experiments on optical quenching of the photohysteresis effect directly indicate that the carriers localized in the traps play the predominant role in photoferroelectric phenomena.

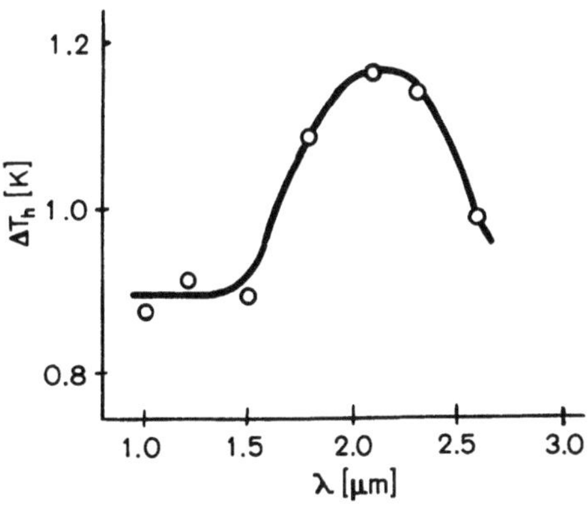

Fig. 5.7. Spectral distribution of quenching of photohysteresis effect in SbSI

In SbSI crystals the photostimulated shift of the Curie point and the photohysteresis effect are similar in the order of magnitude and cannot be clearly differentiated. The importance of distinguishing between these phenomena experimentally has urged the authors of [5.12] to investigate them in solid solutions of $SbSI_x Br_{1-x}$ showing low-temperature phase transitions. Illumination of these crystals at the photoconductivity maximum at a low temperature results in metastable recharging of the levels, which is preserved in the dark for a long time; the latter depends on the temperature and the activation energy of the levels. According to the conclusions of Sect. 2.3 this must lead to a metastable shift of the Curie point which is stable during the entire time of optical recharging of the levels ('frozen' shift). It is significant that under conditions of a frozen shift of the Curie temperature no photohysteresis effect must be observed (if it is due to a change in screening energy in accordance with the mechanism proposed above), since there is no thermal exchange between the deep traps and the band at $u/kT \gg 1$.

Crystals of the composition $SbSI_{0.35} Br_{0.65}$ were investigated in [5.12]. These showed a first-order low-temperature phase transition at $T_1 \simeq 160$ K. Experiment confirmed the previous assumptions. The phase transition in the crystal was observed from the temperature variation of the dielectric constant, measured in the dark on heating and cooling. After that the crystal was heated to room temperature and again cooled under continuous intrinsic illumination to a temperature of 20-25 K above that of the phase transition. Since the crystal was illuminated in the paraelectric phase, no screening effect could be observed.

After switching off the illumination, the specimen was cooled below the phase transition temperature, and the temperature dependence of the dielectric constant was again measured (Fig.5.8), after which the thermostimulated currents for the excited crystal were measured. After illumination both branches

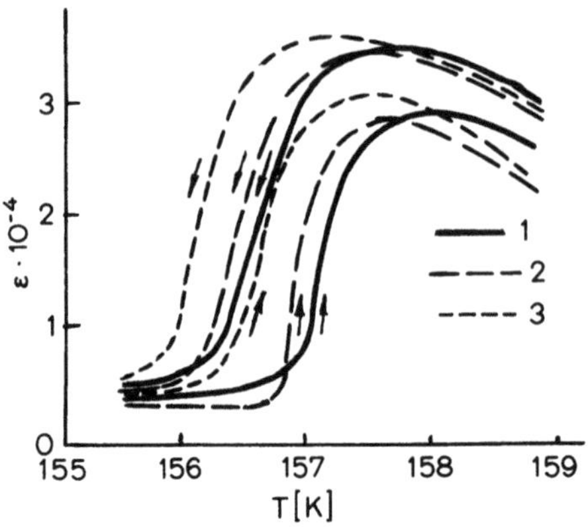

Fig. 5.8. Temperature dependence of dielectric constant for unilluminated (1) and illuminated (2,3) crystal of $SbSI_{0.35}Br_{0.65}$

of the dielectric constant shift toward the low-temperature region, but the temperature hysteresis of the phase transition remains unchanged. In accordance with the above this behavior of the curves $\varepsilon = \varepsilon(T)$ can be interpreted as a shift of the Curie point. The value of the shift is related to the degree of filling of the traps, which is proportional to the area under the thermostimulated current curves. The maximum shift was of the order of ~ 0.5 K. The activation energy of the traps, determined from the initial slope of the thermostimulated current curves [5.13] is ~ 0.6 eV.

The measurements in [5.12] were carried out for several crystals. No photohysteresis effect was observed for any of them. At the same time the frozen shift of the Curie temperature could be observed in many transitions from one phase to the other, provided the temperature did not rise considerably in the course of the experiment. Heating to room temperature restored the dark equilibrium value of the Curie temperature. An attempt at deexcitation of the traps with infrared light and quenching the frozen-shift effect failed. The absence of the deexcitation effect may be due to the fact that away from the phase transition the cross section for capture of carriers by traps is relatively large, whereas the photon capture cross section is small.
The influence of illumination of the Curie temperature in $BaTiO_3$ has been studied using dielectric, pyroelectric, and optical observations [5.14,15]. In dielectric measurements the phase transition temperature was recorded from the maximum of the dielectric constant ε, measured at a frequency of 1 kHz in the [001] direction. Illumination of the crystal with intrinsic light was effected through semitransparent gold electrodes sputtered upon the (001) face.

The results of the dielectric measurements for a pure single crystal of $BaTiO_3$ grown by the Remeika method are presented in Fig. 5.9. In this figure

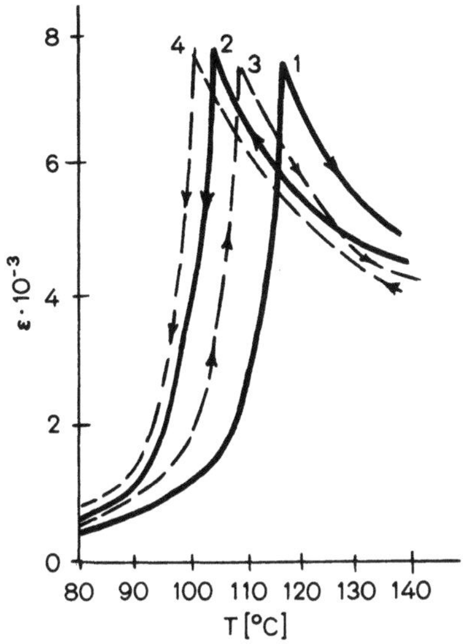

Fig. 5.9. Influence of illumination on phase transition temperature in $BaTiO_3$. 1,2:heating and cooling of crystal in the dark; 3,4:heating and cooling of crystal under illumination

curves 1 and 2 can be used for determining the dark temperature of the phase transition on heating and cooling of the crystal, respectively. Curves 3 and 4 were taken on illumination of the crystal during the phase transition. Thus, in conformity with theory, the Curie point of $BaTiO_3$ is lowered by the influence of nonequilibrium carriers. The dependence of the value of the shift on the light intensity shows saturation. The value of the shift is $\Delta T_N \simeq -7K$, -3.7 K determined from the branch $\varepsilon = \varepsilon(T)$, which corresponds to heating and cooling respectively, Illumination reduces the temperature hysteresis of the phase transition, and $\Delta T_{hN} \simeq -3.3$ K.

In distinction to SbSI, the photostimulated shift of the phase transition in $BaTiO_3$ shows a considerable relaxation time. The equilibrium value of the phase transition temperature is restored within about several hours. Such a slow relaxation points to localization of nonequilibrium carriers at sufficiently deep levels and to the possibility of multiple trapping.

The same measurements show fairly good agreement with (2.42), indicating that a decrease in the temperature hysteresis of the phase transition is accompanied by a photodielectric effect in $BaTiO_3$. The photodielectric effect was also investigated for SbSI [5.4]. It should only be remembered that another possible mechanism responsible for the photodielectric effect in $BaTiO_3$ and SbSI may consist in the influence of intrinsic illumination on the domain structure (the photodomain effect, see Chap.8).

Determination of the coefficients a, b, and c in the expansion (2.46) from the shape of the intrinsic absorption edge of $BaTiO_3$ makes it possible to compare the value of the photostimulated shift of the phase transition temperature and the photohysteresis effect with the thermodynamic calculation. Substituting, for $BaTiO_3$, the values $a \simeq +0.8 \times 10^{-23}$, $b \simeq +2.6 \times 10^{-33}$, $C \simeq 0$ in absolute CGSE units [1.14], $\beta \simeq -7 \times 10^{-13}$ CGSE, and $C \simeq 3 \times 10^5$ K [5.1] in (2.36,39) and, taking the experimental values $\Delta T_N \simeq -5$ K and $\Delta T_{hN}/\Delta T_h \simeq -0.3$, we obtain, from both (2.36) and (2.39), the same value of the concentration of nonequilibrium carriers to an order of the magnitude: $N \simeq 10^{19}$ cm^{-3}.

This result suggests two conclusions. Firstly, photoferroelectric phenomena in $BaTiO_3$, as well as in other ferroelectrics, are associated with carriers localized in traps. Secondly, the saturation values of ΔT_{hN} and ΔT_N evidently indicate extreme filling of the traps in $BaTiO_3$. The level concentration value, $\sim 10^{19}$ cm^{-3}, appears to be genuine for such a high-resistance semiconductor as barium titanate.

Unfortunately, it has not yet been possible to determine the concentration of traps directly by the thermostimulated-current method, as was done for SbSI, probably because of the low values of carrier mobility in $BaTiO_3$. Estimation of the depth and concentration of traps in unreduced crystals of $BaTiO_3$ was carried out in [5.16] by measuring the currents limited by the space charge. For the concentration of traps a value of $N \simeq 3 \times 10^{18}$ cm^{-3} was obtained, and for the activation energy of traps, $u \simeq 0.35$ eV. This concentration of traps obtained is close to our estimate.

The photoferroelectric phenomena described here were not associated with the anomalous photovoltaic effect in $BaTiO_3$ and SbSI (Chap.6). In the first place, the fields generated in the photovoltaic effect shift the Curie point toward the higher temperatures. Secondly, these fields are small at the temperature of the phase transition in SbSI and $BaTiO_3$.

5.3 Influence of Electron Excitations on Spontaneous Polarization

Experimental observation of the optical excitation of Cr^{3+} and Cu^{2+} ions in $LiTaO_3$ and $LiNbO_3$ shows that there is a change in the spontaneous polarization associated with the change in the dipole moment of the impurity ([5.17, 18]; see also Sect.3.4). The kinetics of this photoferroelectric effect depends on the lifetime of the excited states, τ_e, which can be independently determined from the luminescence kinetics.

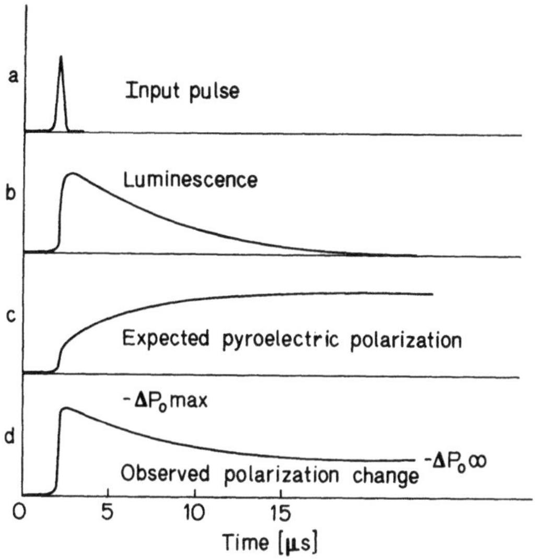

Fig. 5.10. Influence of photoactive illumination on spontaneous polarization [5.12]. (a) exciting pulse; (b) luminescence; (c) expected pyroeffect; (d) observed change in spontaneous polarization

Figure 5.10 shows the kinetics of the change in spontaneous polarization, ΔP_0(d), and also in luminescence (b) on excitation of $LiNbO_3$:Cr by a light pulse (a). The same figure shows the possible change in ΔP_0 associated with the pyroeffect, provided that the entire light energy absorbed is expended on the heating of the crystal (c). A comparison of (c) and (d), as well as the agreement between the experimentally observed kinetics of ΔP_0 and the luminescence kinetics, suggests that the change in spontaneous polarization is associated with the change in the dipole moment of the Cr^{3+} ion when excited. Under this assumption the kinetics of the change in spontaneous polarization satisfies the equation [5.18]

$$\frac{\partial P_0}{\partial t} + \frac{1}{\tau_e} P_0 = \frac{\alpha\Delta\mu_0}{h\nu} I(t) + \alpha\,(1 - e)\,\frac{p}{C}\,\frac{1}{\tau_e} \int I dt \quad , \tag{5.7}$$

where $\Delta\mu_0$ is the change in dipole moment per photon, $h\nu$ is the photon energy, α is the absorption coefficient, I is the light intensity, p is the pyrocoefficient, C is the heat capacity, and e is the radiative recombination probability.

Experiments in which $LiTaO_3:Cr^{3+}$ and $LiNbO_3:Cr^{3+}$ crystals are excited at various temperatures in the Cr^{3+} band with the aid of a dye-based laser show that (5.7) holds well for this case [5.18]. By measuring $\Delta P_{0\,max}$ and $\Delta P_{0\,\infty}$ (Fig.5.10d) and by independent estimation of p/C and e, values of

$\Delta\mu_0 \simeq 3 \times 10^{-18}$ cm^2 V^{-1} s^{-1}, and 4.8×10^{-18} cm^2 V^{-1} s^{-1}, respectively, were found for $LiNbO_3:Cr^{3+}$ and $LiTaO_3:Cr^{3+}$. For a concentration $N \simeq 8 \times 10^{19}$ cm^{-3} of Cr^{3+} ions, $N\Delta\mu_0 = \Delta P_0 \simeq 10^{-7}$ C/cm^2 or $\Delta P_0/P_0 \simeq 10^{-3}$. These values of $\Delta\mu_0$ are independent of temperature up to room temperature. At the same time excitation in Nd^{3+} of the $G_{5/2}$ level in $LiNbO_3:Nd^{3+}$ did not yield a perceptible value of $\Delta\mu_0$, and thus no photoferroelectric effect was observed. (Instead, the ordinary pyroeffect, shown in Fig.5.10c appeared.) In $LiNbO_3$:Fe this effect could not be observed at all, since a strong photovoltaic effect due to the transition of electrons from the Fe^{2+} levels to the conduction band was superposed (see Chap.6).

This photoferroelectric effect has been used for recording short (picosecond) light pulses [5.19]. Picosecond laser pulses at a wavelength of 1.06 μm were recorded by this method using the crystal $LiTaO_3:Cu^{2+}$.

5.4 Photodeformation Effect

The photodeformation effect, or influence of nonequilibrium electrons on the deformation of a ferroelectric, is another consequence of the thermodynamic arguments of Sect. 2.3. According to (2.40) this effect may be associated with the influence of nonequilibrium carriers on the spontaneous polarization through the term $(\Delta P_0/P_0)^2$ and on the electrostriction coefficient $\Delta\nu/\nu$, and also with the lattice deformation due to the pressure of nonequilibrium carriers and described by the term $N(\partial\tilde{E}/\partial\sigma)$, where $\tilde{E}$ is the energy of the excited carriers, and σ is the corresponding component of the mechanical stress tensor. The photodeformation effect may also be associated with the anomalous photovoltaic effect (Chap.6) if the generated photovoltaic field is so high that it causes perceptible deformation due to the inverse piezo-electric effect.

The photodeformation effect was discovered and investigated in SbSI [5.20,21]. Needle-shaped crystals of SbSI were placed in the chamber of a dilatometer, which made it possible to measure and stabilize the temperature, illuminate the crystal with a monochromatic light (in particular, in the region of its intrinsic photosensitivity), and measure the deformation of the crystal in the direction of the spontaneous polarization (along the needle axis).

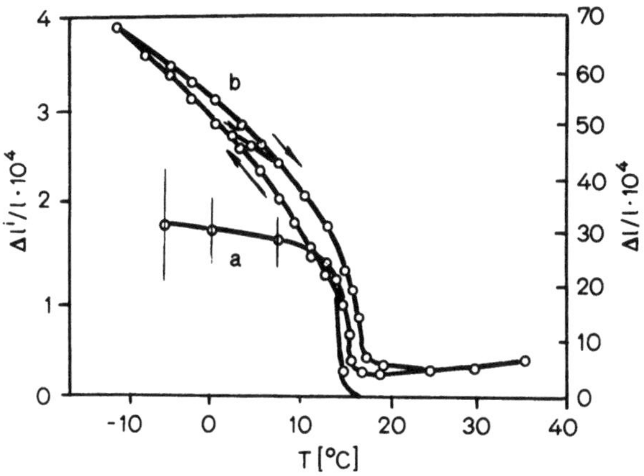

Fig. 5.11. Temperature dependence of deformation of SbSI under illumination (a) and in the dark (b)

Figure 5.11 shows the temperature dependence of the deformation (elongation) of polydomain single crystals of SbSI as a result of illumination with intrinsic light. The measurements were made as follows. The crystal was heated to a temperature of 50°C and cooled slowly for several hours in a darkened chamber, below the phase transition temperature, $T_1 \simeq 19°C$. When a given temperature was reached, the light was switched on, and the deformation $\Delta\ell$ was measured.

The spectral distribution of this effect coincides with the spectral dependence of the photocurrent in SbSI. The photodeformation effect has a positive sign, reduces to zero near the Curie point, and is not observed in the paraelectric phase. If the crystal was first polarized by an external field greater than the coercive field, then either the illumination did not cause additional deformation, or the latter was small. Curve b in Fig. 4.17 demonstrates the temperature dependence of the crystal deformation near the phase transition in the darkened chamber. The arrows indicate the direction of temperature variation. Switching off the light stops the photodeformation process, after which one observes a very slow relaxation to the dark equilibrium value, which was practically never attained in the experiments.

According to the data of Fig. 4.17 the experimental value of the photodeformation effect is $u_N/u = \Delta\ell/\ell \simeq 10^{-4}$. It is interesting to compare the thermodynamic calculation in Sect. 2.3. According to (2.40) the contribution of the spontaneous polarization to the spontaneous photodeformation effect is determined by the term $(\Delta P_0/P_0)^2$. The influence of electron excitations on the spontaneous polarization of a ferroelectric was discussed in the preceding section. In particular, it was shown from independent measurement that the maximum observed value of $\Delta P_0/P_0$ is 10^{-3} (for $LiNbO_3$:Cr^{3+} with

$N \simeq 8 \times 10^{19}$ cm^{-3}). Thus, $(\Delta P_0/P_0)^2_{max} \simeq 10^{-6}$ and hence the contribution of the change in spontaneous polarization to the photodeformation effect may be neglected.

Investigation of the influence of hydrostatis pressure on the energy gap E_g and the piezoelectric effect in SbSI [5.22,23] suggests that over a wide range of temperatures in the ferroelectric and paraelectric phases, $(\partial E_g/\partial p)_T$ and $(\partial E_a/\partial p)_T$ for SbSI vary between 10^{-5} and 10^{-6} eV/atm (E_a is the activation energy of the donor levels). Thus, for a concentration of nonequilibrium carriers in the traps of $N \simeq 10^{18}$ cm^{-7}, the term $N\tilde{E}'$ may vary between 10^{-5} and 10^{-6} for SbSI. So, the contribution of the third term in (2.40) is also insignificant. Although no data on the influence of the electrons on the electrostriction coefficient ν for SbSI are available in the literature, all other experimental results (e.g. the photodeformation kinetics) suggest that the photodeformation effect cannot be estimated from (2.40), and is due to a change in the domain structure on illumination (the photodomain effect). Both the relaxation time of the effect and the temperature hysteresis corroborate this conclusion.

It has already been mentioned above that photoferroelectric phenomena may be based either on the mechanism of optical recharging of levels in a neutral crystal or on the mechanism of screening of the spontaneous polarization by nonequilibrium carriers. Naturally, the thermodynamic approach developed in Sect. 2.3 does not enable us to separate these two possible mechanisms. Nevertheless, although the photodomain effect is one of the screening phenomena, the unsuitability of (2.40) for describing the photodeformation of SbSI crystals is due to the fact that this relation was obtained for single-domain crystal. A change in domain structure on illumination results in additional deformation of the crystal because of the deformation of the domain walls. This mechanism evidently makes the decisive contribution to the photodeformation of SbSI.

The photodeformation effect in SbSI was observed in temperature range (see Fig.5.11) where the photovoltaic field can be neglected in view of the high electrical conductivity (Chap.6). In $LiNbO_3$:Fe, on the contrary, the photodeformation effect was observed simultaneously with the photorefractive effect (Chap.1), and it was shown that both of them are due to the anomalous photovoltaic effect [5.24]. In [5.24], local photodeformation of the crystal, $\Delta\ell/\ell \simeq 10^{-3}$, in the c-direction was observed by exciting $LiNbO_3$:Fe with the aid of a 20 mW He-Ne laser. The photodeformation was estimated from an X-ray topogram (the Bormann effect). Like the photorefractive effect, the photodeformation showed a slow thermal relaxation (months) and disappeared completely on irradiation of the whole crystal with ultraviolet light ('erasing' the photorefractive effect, Chap.7).

5.5 Photoinduced Rayleigh Scattering in $BaTiO_3$

The existence of the photohysteresis effect in $BaTiO_3$ and other photoferroelectrics and a reduction in coefficient $|\beta_N| < |\beta|$ in the expansion (2.14) on illumination support the conclusion in Sect. 2.3 that nonequilibrium carriers change the nature of the phase transition, bringing it closer to the critical Curie point at which $\beta = 0$. This, in turn, must result in an appreciable increase of Rayleigh scattering near the phase transition in the ferroelectric. Indeed, GINZBURG [5.25] has shown that the intensity of light scattered by polarization fluctuations near the phase transition is proportional to β^{-1} and must therefore (by analogy with critical opalescence) show a sharp maximum at the Curie point, which is close to the critical point. Therefore it should be expected that in $BaTiO_3$ and other photoferroelectrics an increase in concentration of nonequilibrium electrons intensifies Rayleigh scattering at the Curie point ('photoinduced Rayleigh scattering').

Photoinduced Rayleigh scattering was discovered in $BaTiO_3:Fe^{3+}$ on photoexcitation of electrons from the iron levels and also in the intrinsic absorption region [5.26-28]. a-domain crystals grown by the Remeika method were investigated. Scattering of the light of a He-Ne laser without additional illumination from a xenon lamp in the intrinsic or extrinsic region (photoinduced scattering) was studied. It was found that illumination in the intrinsic or extrinsic region ($\lambda \simeq 547$ mμ) increases the intensity of the scattered light everywhere in the ferroelectric region, but near the phase transition there is a sharply defined photoinduced-scattering maximum (Fig.5.12). In dark scattering, the light scattered by the crystal is not polarized (more precisely, its polarization is determined by the polarization of the incident light). In photoinduced scattering, the light scattered by the crystal is polarized in the direction of spontaneous polarization. On switching of a crystal in the external field, when the spontaneous polarization reverses its direction in the a plane, the direction of the polarization of the scattered light changes accordingly. In [5.27,28], the spectral composition of photoinduced

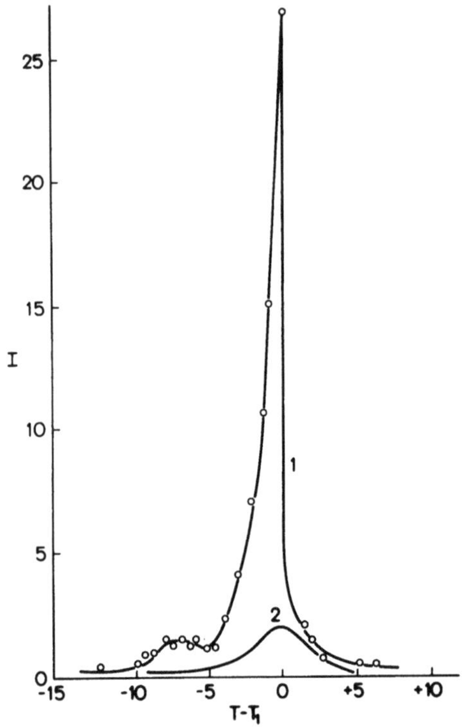

Fig. 5.12. Photoinduced light scattering in $BaTiO_3$:Fe [5.27,28]. Dependence of intensity of scattered light in arbitrary units on $T - T_1$:1, photoinduced scattering; 2, dark scattering

Rayleigh scattered light in $BaTiO_3$:Fe was investigated by Fourier spectroscopy. It was shown that the photoinduced scattering is of a dynamic nature, and as the Curie point is approached, both the scattering intensity and the spectral half-width γ increase (Fig.5.13). According to Fig. 5.13, the maximum value of 2γ at the Curie point does not exceed 40 kHz, and 5 K away from the Curie point it drops below 10 kHz, i.e. below the accuracy of the measurements. Thus, the photoinduced scattering in $BaTiO_3$:Fe reveals a dynamic character only in a narrow temperature range near the phase transition. Away from the Curie point the photoinduced scattering (just like dark scattering) is static.

The existence of polarized photoinduced scattering points to an inhomogeneous distribution of the spontaneous polarization in an excited photoferroelectric. This may be due to the generation of photoinduced polarization fluctuations, which in turn may be caused by the impurity Jahn-Teller effect (Sect.3.3) or by optical recharging of the centers and a change in their dipole moment (Sect.3.4). A mechanism associated with the formation of phasons stabilized by nonequililibrium electrons that are captured by heterogeneous fluctuations is not excluded, either (Sect.3.5). This would explain, among

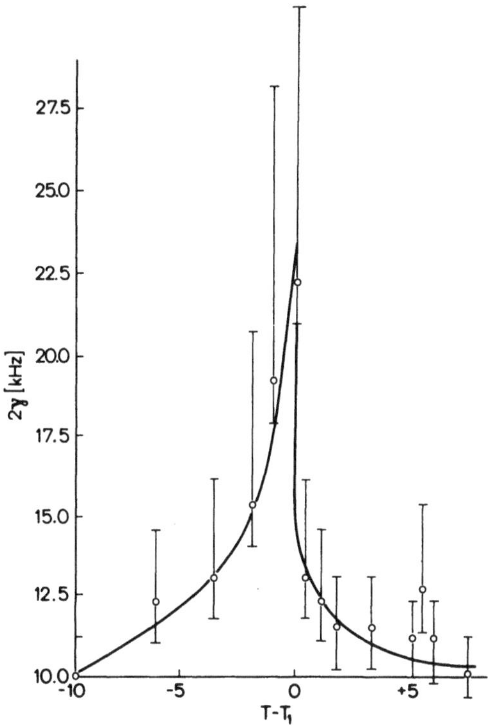

Fig. 5.13. Dependence of spectral width of photoinduced scattering 2γ on $T - T_1$ in $BaTiO_3$:Fe [5.27,28]

other things, the abrupt increase in photoinduced scattering at the Curie point and its dynamic nature in the phase transition region.

If we assume that the phason lifetime is restricted by the electron localization time at the level of a phason with an activation energy u_0 and a recombination cross section $S = \pi R_0^2$, then the spectral width of light scattering on phasons is given by

$$2\gamma = k_0 = 2\beta N_c \exp\left(-\frac{u_0}{kT}\right) \quad , \tag{5.8}$$

where $\beta = Sv$ is the coefficient of recombination, $v = (2kT/m)^{1/2}$ is the thermal velocity of an electron, N_c is the density of states in the band, k_0 is the probability of a thermal transition of an electron from the phason level to the band, and the activation energy u_0 and the phason radius R_0 satisfy (3.37). Substituting $\varphi \equiv F = 0$, $\sigma \simeq 7$ erg$\cdot$cm^{-2}, $m = m_0$ (m_0 is the mass of the free electron), $v \simeq 1.1 \times 10^7$ cm$\cdot$sec^{-1} into (3.37,5.8) for $BaTiO_3$ at the Curie point, we have $u_0 \simeq 0.63$ eV, $R_0 \simeq 7$ Å, and $2\gamma \simeq 34$ kHz. Thus, the spectral width at the Curie point agrees with the experimental value. As we recede from the Curie point, $\varphi \equiv F$ increases and according to (3.37), the phason activation energy u_0 increases too, which leads, in accordance with (5.8), to a sharp drop of 2γ in a narrow temperature range near the Curie point. Thus, with an increase in the activation energy of the phason level (with the distance from the Curie point) the scattering changes from being dynamic to static. The maximum scattering intensity at the Curie point in this case follows directly from the temperature dependence of the phason concentration (3.41).

As has already been indicated in Sect. 3.5 for $BaTiO_3$ and other ferroelectrics the phason shape is ellipsoidal, elongated in the direction of spontaneous polarization; this explains, in particular, the polarized nature of photoinduced scattering. Quantitative comparison of theory with experiment, however, only make sense when (3.35) takes into account the energy of the depolarizing field. It is possible that under conditions of nonequilibrium conductivity the depolarizing field is effectively screened, and $\ell_D \ll R_{||}$, where $R_{||}$ is the phason dimension in the direction of spontaneous polarization.

6. The Anomalous Photovoltaic Effect in Ferroelectrics

It is well known that the photovoltage in semiconductors, whatever its nature, cannot exceed the energy gap, which is several volts. For instance, in a homogeneous semiconductor the Dember (diffusion) photovoltage does not exceed the following value [3.6] for an infinitely large intensity of the exciting light:

$$V = \frac{kT}{q}\,\ell n\,\frac{n_1}{n_0} << \frac{kT}{q}\,\ell n\,\frac{N_c}{n_0} = E_g \quad . \tag{6.1}$$

Here, E_g is the energy gap of the semiconductor, n_1 and n_0, respectively, are the nonequilibrium and equilibrium carrier concentrations, and N_c is the density of states. Another example is the photovoltage resulting from the illumination of a p-n junction [5.7]

$$V \lesssim \frac{kT}{q}\,\ell n\,\frac{n_n p_p}{n_0^2} = E_{Fn} - E_{Fp} \lesssim E_g \quad , \tag{6.2}$$

which does not exceed E_g, either. Here, n_n and p_p are, respectively, the concentration of electrons in the n region and that of holes in the p region, E_{Fn} and E_{Fp} are the Fermi energies in the n and p regions.

Until recently, the only exception from this rule were semiconductor textures which have anomalously high photovoltages (the APV effect) due to the composition of the elementary Dember photo e.m.f. (6.1), or elementary photo e.m.f. (6.2) which develop on separate p-n junctions of the texture [6.1]. In such textures consisting of evaporation layers of CdTe, Si, Ge, GaAs, PbS, etc. the photovoltages may be as high as several hundred volts per centimeter length in the direction of the composition of the elementary photo e.m.f. (6.1) or (6.2).

Recently, the APV effect has been observed in a number of homogeneous single crystals of ferroelectrics [6.2-5]. It was found that on uniform illumination of a homogeneous ferroelectric with open electrodes, photovoltages of the order of 10^3 - 10^5 V arise in the intrinsic or extrinsic region of the spectrum; these voltages thus exceed the value of E_g by two to four orders of magnitude. The APV effect in a ferroelectric is observed only in the direction of the spontaneous polarization P_0 of the crystal and disappears in the paraelectric region; the value of the photovoltage is proportional to the crystal length in the direction of P_0. It was also shown that the effect of illumination on the optical birefringence of a ferroelectric, or the photorefractive (PR) effect (see Chap.7) and thus the recording of volume phase holograms in ferroelectrics based on the PR effect is due to the APV effect (at least for all the crystals studied so far) [6.6-8]. However, other mechanisms may also be responsible for the PR effect [6.9]. This will be discussed more thoroughly in the next chapter.

When a homogeneous short-circuited ferroelectric is uniformly illuminated, a stationary current passes through it [6.10]. This photovoltaic current J flows in the direction of spontaneous polarization P_0 and is directly proportional to the light intensity I. The photovoltaic current is responsible for the APV effect in a ferroelectric [3.21]. When measuring the photovoltage (in open circuit), a transitional photocurrent

$$J^* = J + (\sigma_d + \sigma_{ph})\mathcal{E} \qquad (6.3)$$

passes throughout the ferroelectric in the direction of P_0 [3.21]. In this equation, $\mathcal{E}$ is the macroscopic electric field formed due to the charging of the crystal capacitance with the photovoltaic current J; σ_d and σ_{ph} are, respectively, the conductivity in darkness and under illumination. The photovoltage V that appears in the crystal in this direction during the Maxwellian relaxation ($J^* = 0$) is

$$V = \tilde{\mathcal{E}}\ell = \frac{J}{\sigma_d+\sigma_{ph}}\ell \quad , \qquad (6.4)$$

where ℓ is the distance between the electrodes. According to (6.4), V is directly proportional to the interelectrode distance and is not limited by the energy gap.

In what follows, the anomalous photovoltaic (AP) effect in ferroelectrics will be used to cover both the stationary photovoltaic current with short-circuited electrodes and the photovoltage $V \gg E_g$ (the APV effect) in open circuit. It should be noted that the APV effect in ferroelectrics may, in principle, be associated, not with the photovoltaic current, but with transitional currents, which are of a different nature. However, for all ferroelectrics investigated so far, this is not the case, and so the analysis of the mechanism of the APV effect amounts to the study of the nature of the photovoltaic current.

6.1 The APV Effect in Ferroelectrics

Photovoltages much greater than the energy gap were first observed in investigations on photoferroelectric phenomena in single crystals of a ferroelectric solid solution of $SbSI_{0.35}Br_{0.65}$ [6.3], and then in single crystals of $BaTiO_3$ [6.4].

Rhombic single crystals of $SbSI_{0.35}Br_{0.65}$ below the temperature of a first-order phase transition $T_1 = 170$ K, belong to the symmetry point group 2m. Their ferroelectric properties are characterized by the following parameters: spontaneous polarization $P_0 = 21 \times 10^{-6}$ C/cm^2, the Curie-Weiss constant $C = 2 \times 10^5$ K, and the Curie-Weiss temperature $T_0 = 165$ K. In the ferroelectric region, near the Curie point, a ferroelectric is a dielectric with a dark conductivity of $\sigma = 10^{-14}\ \Omega^{-1}\ \mathrm{cm}^{-1}$; the spectral maximum of photoconduction is near 550 nm.

The measurements in [6.3] were made in the following sequence. A previously polarized single-domain crystal was illuminated in the ferroelectric phase with light in the region of maximum photosensitivity. On illumination, a short-circuit photocurrent was observed in the [001] direction. Its stationary or transitional nature was neglected. On switching off the light, the crystal was heated above the Curie point, which was recorded from the pyroelectric current maximum. It was found that illumination of the crystal shifts the Curie point towards higher temperatures and that this shift, ΔT_1, is due to the internal field $\tilde{\mathscr{E}}$ of space charges. The field $\tilde{\mathscr{E}}$ was estimated by a method widely used for photoelectrics [6.11]. The crystal was again illuminated in the paraelectric phase. Then the depolarization photocurrent was observed; by integrating this current the space-charge density Q_F was determined, and the internal field $\mathscr{E}$ was estimated from

$$\tilde{\mathscr{E}} = 4\pi \frac{Q_F}{\varepsilon} \quad . \qquad (6.5)$$

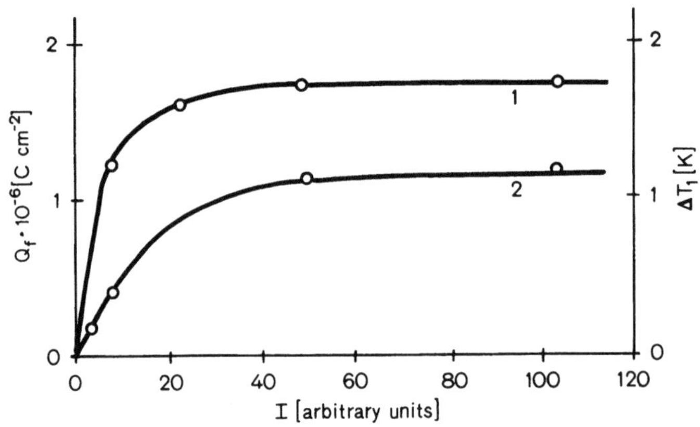

Fig. 6.1. APV effect in $SbSI_{0.35}Br_{0.65}$: Dependence of Q_F (1) and ΔT_1 (2) on light intensity

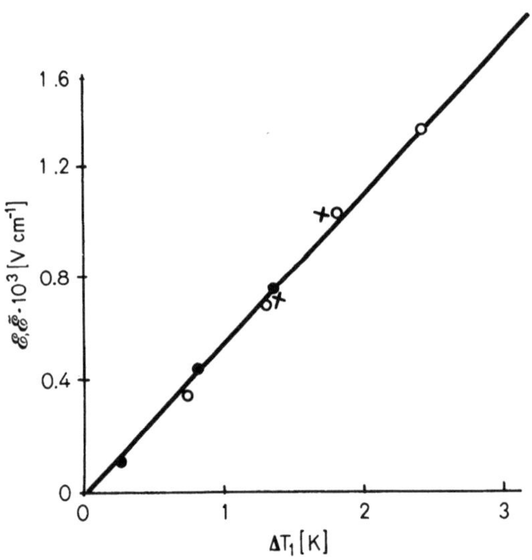

Fig. 6.2. APV effect in $SbSI_{0.35}Br_{0.65}$: shift of phase transition temperature with field, as determined from three groups of independent measurements

Figure 6.1 shows the dependence of Q_F (curve 1) and the Curie-point shift ΔT_1 (curve 2) on the light intensity. For the saturation charge $Q_F = 1.9 \times 10^{-6}$ C/cm^2 and $\varepsilon = 2 \times 10^3$, (6.5) gives for the field $\tilde{\mathscr{E}} = 10.5 \times 10^3$ V/cm or $V = \tilde{\mathscr{E}} \ell \simeq 3 \times 10^3$ V. Figure 6.2 exhibits the linear dependence of the shift of the Curie point ΔT_1 on the field $\tilde{\mathscr{E}}$ obtained thus [6.3]. It is in good agreement with data obtained by applying the external field $\mathscr{E}$ to the crystal in the [001] direction. Finally, both these dependences lead to the value of the coefficient $dT_1/d\mathscr{E} \simeq 0.8 \times 10^{-3}$ K·cm/V, which is in good agreement with the theoretical value

$$\frac{dT_1}{d\mathscr{E}} = \frac{C}{4\pi}\left(-\frac{4}{3}\frac{\gamma}{\beta}\right)^{1/2} , \qquad (6.6)$$

where the Landau-Ginsburg-Devonshire coefficients β and γ were determined from the expansion of the free energy of the crystal in powers of polarization.

The APV effect was similarly observed in single crystals of $BaTiO_3$ [6.4]. c-single-domain crystals of thickness up to 100 μm in the c direction were investigated. After illumination of the crystal in the ferroelectric phase, the Curie-point shift was measured by taking the temperature dependence of the dielectric constant ε [Fig.6.3, curves 3,4]. A comparison with the initial equilibrium values of the phase transition temperature determined from curves 1,2 shows an upward shift of the Curie point with $\Delta T_1 = 4.8$ K. In the dark, the phase-transition point returns to the initial equilibrium value within about an hour. Re-illumination of the crystal in the paraelectric phase depolarizes the crystal and thus brings the Curie temperature back to its equilibrium value. The internal field responsible for the upward shift of the phase transition was estimated by the method described above. For the case presented in Fig. 6.3, $\tilde{\mathscr{E}} = 4 \times 10^3$ V/cm giving a photovoltage V of about 40 volts, which is much greater than E_g. Hence $dT_1/d\tilde{\mathscr{E}} \simeq 1.2 \times 10^{-3}$ K·cm/V, which is close to the coefficient of the shift of the

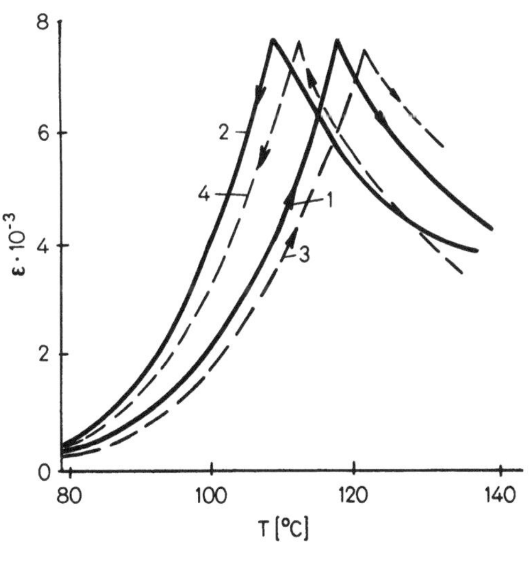

Fig. 6.3. APV effect in BaTiO3: effect of illumination of the ferroelectric phase on Curie point of BaTiO3

phase-transition point in $BaTiO_3$ under the effect of the external electric field, $dT_1/d\mathcal{E} \simeq +(1-1.2)\times 10^{-3}$ K·cm/V [1.13].

With an increase in light intensity the photovoltage and the shift of the Curie point increase. As the pre-polarization field increases, so does the photovoltage, reaching saturation at fields corresponding to the saturation of polarization and pyrocharge. In both cases described the photovoltage and the shift of the Curie point showed a slow dark relaxation due to the localization of nonequilibrium carriers in relatively deep traps.

The APV effect has since been observed in lithium niobate [6.2], in a number of other ferroelectric niobates [6.7], and in the ferroelectric SbSI, and was investigated more thoroughly in barium titanate [6.12-14].

The kinetics of the APV effect was studied for single crystals of $LiNbO_3$:Fe (point group 3m), $KNbO_3$:Fe (2m), and antimony orthoniobate $SbNbO_4$ (2m) [6.5]. The crystals were illuminated with an argon laser at a wavelength of 488 nm in the energy density range up to 500 mW/cm^2. The photovoltage V was measured in the direction of spontaneous polarization [001], for which purpose the electrodes were connected to the input capacitance of an electrostatic voltmeter. Separate observations showed that the photo e.m.f. in the perpendicular direction $V < E_g$. Simultaneously with the measurement of the photo e.m.f. in lithium niobate, observations of the PR effect were made. The PR effect was measured for the $KNbO_3$ crystals of [6.5], but the $SbNbO_4$ crystals were optically inhomogeneous, preventing measurement of the PR effect in them [6.15].

Figure 6.4a,b illustrates the experimental results for a single crystal of lithium niobate with an iron impurity. Crystals of $LiNbO_3$:Fe show a photo e.m.f. of $V \simeq 10^3$ volts and more in the [001] direction (Fig.6.4b) on illumination with a light probe in the shape of a point or a strip of ℓ ranging from 1×10^2 to 5×10^{-2} cm parallel to the electrodes. The light probe was focused on the (100) face. Figure 6.4a displays the kinetics of increase in V at the initial stage of illumination for two light intensities (curves 1 and 2). On switching off the light at an arbitrary time t, the photo e.m.f. drops by a value corresponding to the pyrocharge, after which the value of V slowly relaxes in the dark. Curves 1 and 2 in Fig. 6.4a show V less the pyrosignal as a function of t. The curves have two characteristic regions an initial (fast) and a subsequent (slow) increase. A comparison of this curve with the optical damage in the same crystal, $\Delta n = \Delta n(t)$ (the PR effect), indicates their coincidence in the initial, fast section. In the region of the slow increase of $V = V(t)$ the optical damage retains its saturated value, Δn = const.

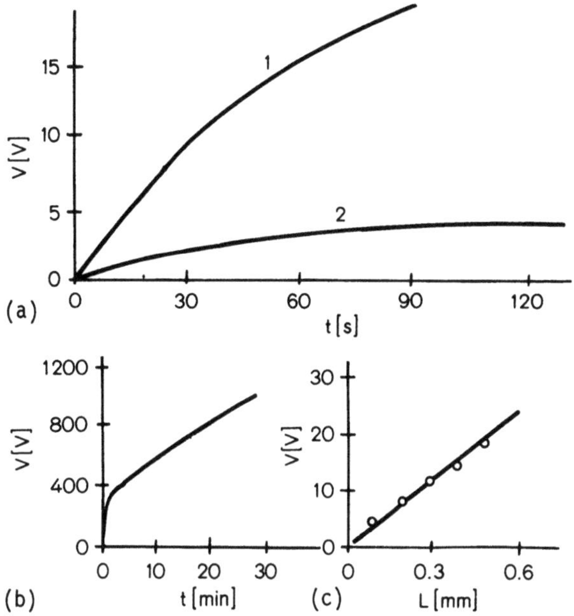

Fig. 6.4. APV effect in lithium niobate. (a) $V = V(t)$ for two light intensities I, W/cm^2: 0.15 (1), 0.02 (2); (b) $V = V(t)$ for longer exposures and higher intensity; (c) dependence of $V(\tau_1)$ on L_1

These data are close to the relaxation times Δn obtained for $LiNbO_3$:Fe by a different technique [6.16]. The values of $V(\tau_1)$ depend linearly on the width of the light probe, L (Fig.6.4b), and also depend on the light intensity I.

The APV effect in lithium niobate is often accompanied by a reversible electrical breakdown, which is manifested as periodic jumps of the photovoltage V and the corresponding periodic changes in optical damage, Δn. This phenomenon may be associated with impact ionization in a strong field $\mathscr{E}$ in the light probe region, although a domain polarization mechanism is not excluded. The latter mechanism is favored by the fact that at high light intensities, the photovoltage corresponds to a depolarization field $\tilde{\mathscr{E}}$ close to the coercive field (see Table 6.1).

The $SbNbO_4$ crystals were illuminated with a light probe in the direction of spontaneous polarization [001]. The photovoltage V was measured in the same direction. The measurement results are presented in Fig. 6.5a,b,c. On switching on the light, the photovoltage V increases within the time τ_1 to a stationary value $V(\tau_1)$, which depends on the light intensity I (Fig.6.5a). After the light is switched off, $V(\tau_1)$ slowly relaxes. The pyroeffect is insignificant and does not affect the measurement of V. Figure 6.5b reflects the dependence of $V(\tau_1)$ on I, which is close to the law $I^{1/2}$ in the region of low I, and Fig. 6.5c shows that the time τ_1 is inversely proportional to I.

Table 6.1. PR effect Δn and depolarization field $\tilde{\mathcal{E}}$ in APV effect

Crystal	n_1	n_3	$r_{13}[mV^{-1}]$	$r_{33}[mV^{-1}]$	ε	$\ell[cm]$	$V(\tau_1)[v]$	$\tilde{\mathcal{E}}[Vcm^{-1}]$	Δn calculated	Δn experi-mental
$LiNbO_3$	2.29 [6.17]	2.22 [6.17]	8.6×10^{-12} [6.17]	30.8×10^{-12} [6.17]	30 [6.17]	3×10^{-2}	830	28×10^3	3.3×10^{-4}	10^{-3} [6.16]
$KNbO_3$	2.28 [6.18]	2.17 [6.18]	28×10^{-12} [6.18]	64×10^{-12} [6.18]	100 [6.19]	5×10^{-2}	10	200	3×10^{-6}	6×10^{-6} [6.15]
$SbNbO_3$	-	-	$<2 \times 10^{-12}$	$<2 \times 10^{-12}$	240 [6.20]	2×10^{-2}	20	10^3	$<1 \times 10^{-6}$	-

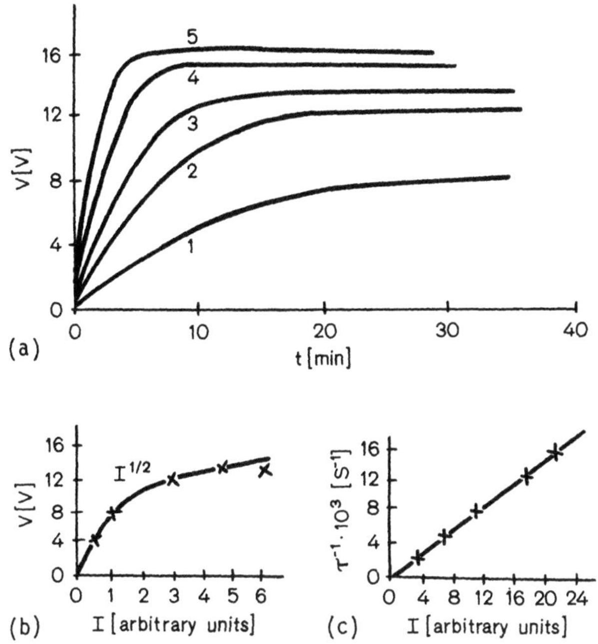

Fig. 6.5. APV effect in antimony orthoniobate: (a) kinetics of APV effect at different light intensities I in W/cm^2: 1 - 0.02 (1), 0.06 (2), 0.12 (3), 0.2 (4), 0.4 (5); dependence of (b') $V(\tau_1)$ on I; dependence of (c) τ_1 on I

In crystals of $KNbO_3$:Fe the photovoltage in the [001] direction is several times the energy gap. The kinetics of the APV and PR effects in potassium niobate are characterized by a common time constant. The photovoltages in $BaTiO_3$ crystals were found to be of the order of 10 volts [6.12-14], agreeing with the earlier investigation [6.4]. In barium titanate, the APV effect depends on the domain structure and is maximal in single-domain high-resistance crystals [5.26].

In conclusion we note that in accordance with (6.4) the value of the APV effect for all the ferroelectrics investigated is inversely proportional to the photoconductivity, and its kinetics are determined by the Maxwell constant. For all the ferroelectrics studied, a simple relationship between the PR and APV effects can be traced, namely that the change in birefringence is the consequence of the linear electro-optical effect due to the field $\tilde{\mathscr{E}}$ induced by the APV effect. This can be seen, for instance, from Table 6.1, where the change in birefringence, Δn, was calculated from the relation

$$\Delta n = \frac{1}{2}\left(n_1^3\, r_{31} - n_3^3\, r_{33}\right)\tilde{\mathscr{E}} \quad , \qquad (6.7)$$

where n_1, and n_3 are the refractive indices; r_{13}, r_{33} are the electro-optical coefficients; and $\tilde{\mathscr{E}} = V/\ell$. The experimental values of Δn for $LiNbO_3$ and $KNbO_3$ agree with the calculated ones.

6.2 Photovoltaic Current with Short-Circuited Electrodes

The stationary photovoltaic current was first observed for $LiNbO_3$:Fe and a ferroelectric solid solution of $Ba_xSr_{1-x}Nb_2O_6$ (SBN) [6.2,10,21]. The relationship between the photovoltaic effect and the PR effect in $LiNbO_3$ was established later [6.6-8,22]. The stationary photovoltaic current and the APV effect were studied independently in lithium niobate in [6.2], where it was shown, using (6.3,4), that the APV effect is due precisely to the stationary photovoltaic current; to explain the latter, a new mechanism of carrier transport in a ferroelectric was proposed. Later on, the AP effect (including the photovoltaic current) was studied in a series of investigations [3.21;6.7,8, 23].

Figure 6.6 shows the kinetics of the photovoltaic current in $LiNbO_3$:Fe upon switching the light on and off [6.21]. Since the Maxwell time in $LiNbO_3$ at T = 300 K does not exceed a few minutes ($\sigma \simeq 10^{-14}\ \Omega^{-1}\ cm^{-1}$, $\varepsilon \simeq 30$), the stationary photovoltaic current observed for tens of minutes could not be the screening photocurrent, as is suggested in [6.16]. Similar results were obtained for an SBN crystal, for which the Maxwell time at T = 300 K does not exceed a few seconds. The kinetics of the transitional maximum

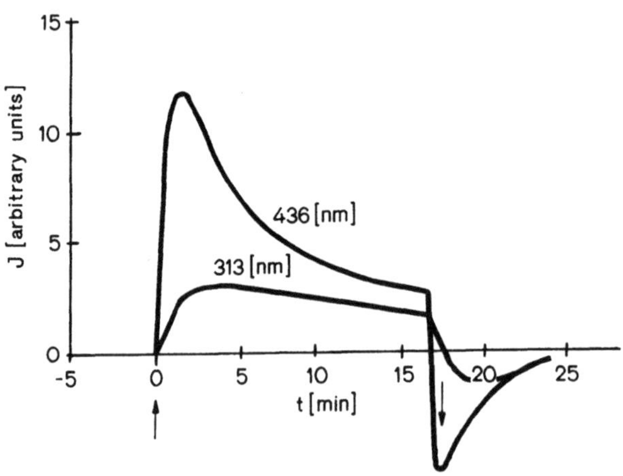

Fig. 6.6. Kinetics of photoresponse in $LiNbO_3$ when measuring photovoltaic current

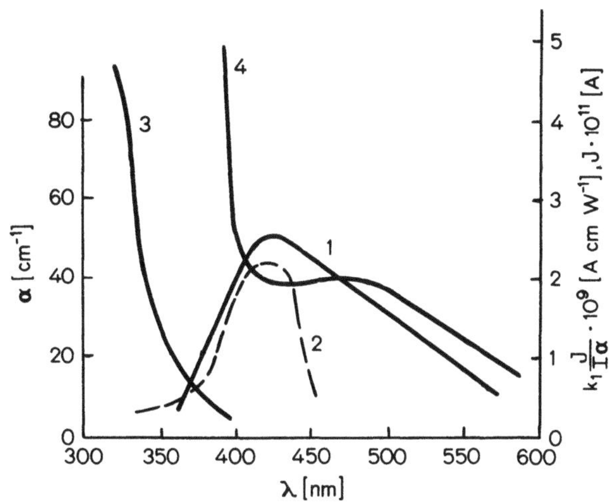

Fig. 6.7. Spectral distribution of absorption (4), photoconductivity (3), and photovoltaic current (1,2)

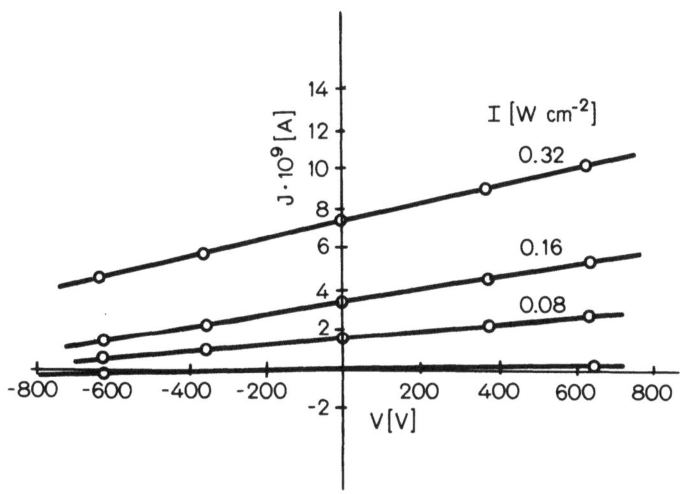

Fig. 6.8. Current-voltage characteristics of $LiNbO_3$:Fe at T = 300 K, $\lambda \simeq 473$ nm and different light intensities [3.21]

upon switching on the light (Fig.6.6) are determined by the Maxwellian relaxation of the photovoltaic current associated with the formation of space charge in the crystal. Therefore the transitional maximum decreases upon successive switching of the light on and off. This makes it possible to use the change in transitional maximum for estimating the space-charge field $\tilde{\mathscr{E}}$. Of course, the pyrocurrent may also be superposed on the transitional maximum. Figure 6.7 exhibits the spectral distribution of the stationary

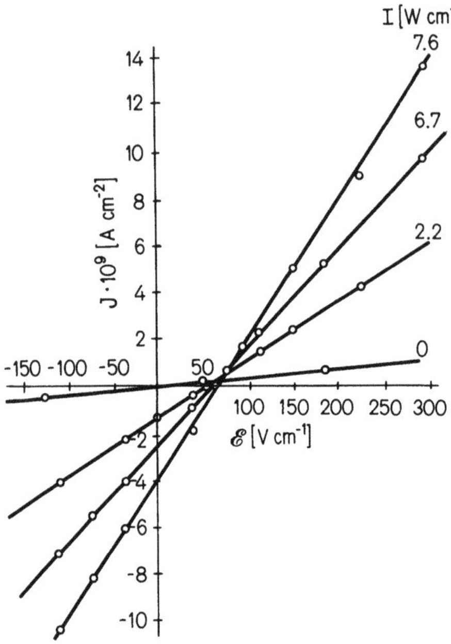

Fig. 6.9. Current-voltage characteristics of $KNbO_3$:Fe at T = 300 K, $\lambda \simeq 488$ nm and different light intensities [6.7]

photovoltaic current in $LiNbO_3$:Fe (curve 1) obtained in [6.41] and later confirmed in [3.21] (curve 2, Fig.6.7). The spectral distribution of the photovoltaic current will be discussed below. Here we shall only note that for $LiNbO_3$:Fe this distribution shows a maximum near λ = 400 nm, which corresponds to the Fe^{2+} band and is not visible (or is only slightly visible) in the photoconduction spectrum (curve 3, Fig.6.7).

In subsequent investigations the effect was studied by taking current-voltage characteristics of the photocurrent over a wide range of light intensities. Fig. 6.8 gives the current-voltage characteristics of the photocurrent in $LiNbO_3$:Fe [3.21]. From (6.3) and (6.4), the photovoltaic current J is measured by intercepts on the current axis, and the photovoltage V or the field $\tilde{\mathcal{E}}$, corresponding to the APV effect, by intercepts on the voltage axis. It can be seen from Fig. 6.8 that for light intensities I in the range of 1 - 10^{-1} W/cm^2 the photovoltaic current varies between 10^{-8} and 10^{-9} A/cm^2, which induces a photovoltage of 10^3 - 10^4 volts in a crystal of length ℓ = 1 cm. Thus, the intensity-current characteristic of the photovoltaic current in $LiNbO_3$:Fe is linear and is given to a good approximation by an expression of the form

$$J[\mathrm{Acm}^{-2}] \simeq 10^{-8} I\ [\mathrm{Wcm}^{-2}] \quad (6.8)$$

In the light intensity range given in Fig. 6.8 the photovoltage V increases linearly with I. According to (6.4) this corresponds to a low photoconductivity $\sigma_{ph} < \sigma_d$ in lithium niobate. In $KNbO_3$:Fe the lifetime of the photocarriers is much larger, and thus for I of the order of $1W/cm^2$, $\sigma_{ph} >> \sigma_d$, and in accordance with (6.4) saturation is observed in the dependence V = V(I). This is illustrated by Fig. 6.9 [3.21]. Note that photovoltages measured by taking the current-voltage characteristics correspond to the APV effect in open circuit. The intensity-current and temperature characteristics of the photovoltaic current J differ drastically from the corresponding characteristics of the photocurrent J_{ph} [6.23].

Figure 6.10 presents the intensity-current characteristics of J and J_{ph} for SBN crystals ($Ba_{0.25}Sr_{0.75}Nb_2O_6$) in the tetragonal phase with point group 4 mm (a), $SbNbO_4$ (b), and $BaTiO_3$ (c) at room temperature. For all the three crystals the intensity-current characteristics of J are linear over the entire range of light intensities I, whereas the photocurrent J_{ph} has a sublinear intensity-current characteristic $J_{ph} \sim I^{1/2}$ ($SbNbO_4$, $BaTiO_3$) or a superlinear section (SBN). GLASS et al. [6.2] obtained an empirical intensity-current characteristic of the photovoltaic current in $LiNbO_3$

$$J = k_1 \alpha I \quad ,$$

where α is the light absorption coefficient, I is the intensity of the exciting light, and k_1 is the Glass constant. On illumination in the band of Fe^{2+} at T = 300 K, $k_1 \simeq (2 - 3) \times 10^{-9}$ A cm/W and is independent of the iron concentration.

Figures 6.11-14 show the temperature dependences of the photovoltaic current J and the photoconductivity σ_{ph} for $BaTiO_3$, SBN, $LiNbO_3$, and SbSI. While σ_{ph} falls off exponentially with decreasing temperature, J increases by a power law. Thus, for barium titanate (Fig.6.11) the temperature dependence J = J(T) was found to be close to $\sim T^{-3.5} (T - T_1)^{1/2}$, where T_1 is the Curie temperature, which corresponds to the temperature dependence of the mobility μ [1.14]. A similar result was obtained for SNB, $LiNbO_3$, and SbSI, where the temperature dependence J = J(T) is well approximated by the power dependence $\sim T^{-3}$ and, hence, according to [6.24] it is close to the temperature dependence of μ. For $LiNbO_3$, Fig. 6.13 gives the experimental values of the mobility μ taken from [6.25], which fit well on the temperature dependence of the photovoltaic current J obtained in [6.23] (the photoconductivity in $LiNbO_3$ is low, and was not measured). Thus, the data of [6.23] indicate, firstly, that the photo-

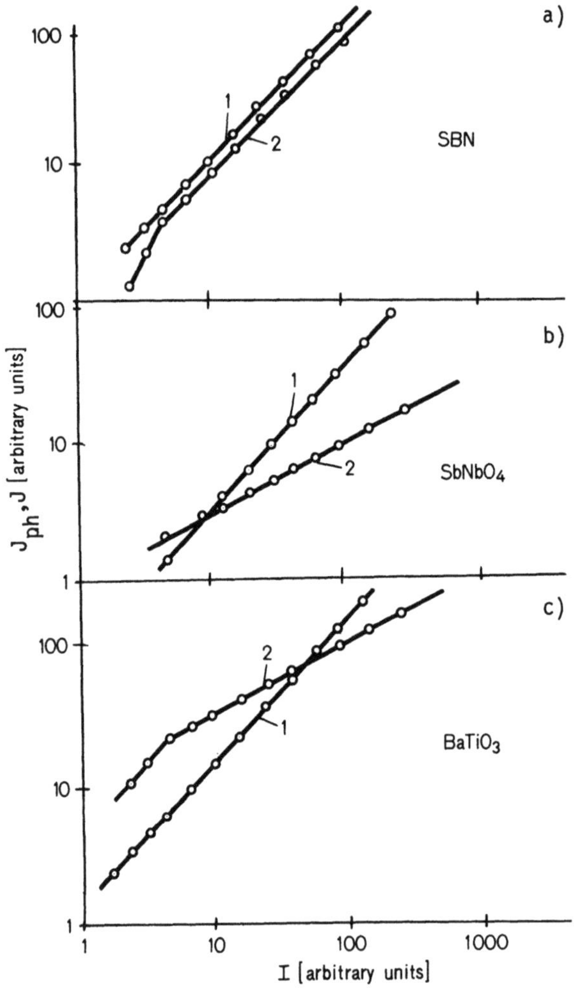

Fig. 6.10. Intensity-current characteristics of J (1) and J_{ph} (2) for SBN, $SbNbO_4$ and $BaTiO_3$

voltaic current and photoconductivity vary with the temperature in opposite directions, and, secondly, that the weak [as compared with $\sigma_{ph} = \sigma_{ph}(T)$] temperature dependence $J = J(T)$ may be associated with that of the mobility.

Whereas the photovoltaic current J only depends slightly on the temperature, the photovoltage V or the corresponding field $\tilde{\mathscr{E}}$ must [in accordance with (5.4)] increase exponentially with decreasing temperature. In fact, $\tilde{\mathscr{E}} = \tilde{\mathscr{E}}(T)$ correlated with the temperature dependence of the electrical conductivity on illumination, $\sigma = \sigma(T)$ and of the photorefractive effect, $\Delta n = \Delta n(T)$. This is illustrated for SNB by Fig. 6.15.

A comparative study into the spectra of the photovoltaic current and photoconductivity was carried out in a number of investigations [6.7,23,26]

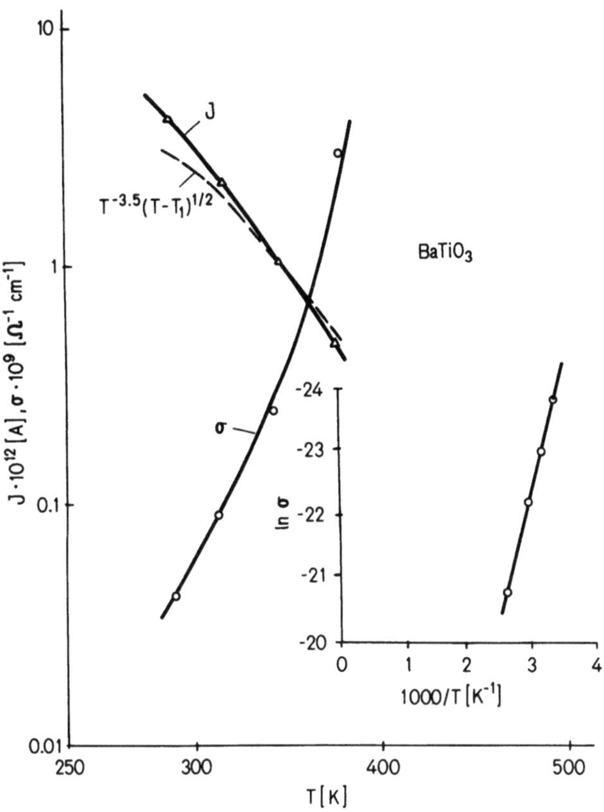

Fig. 6.11. Temperature dependences of J and σ_{ph} for $BaTiO_3$

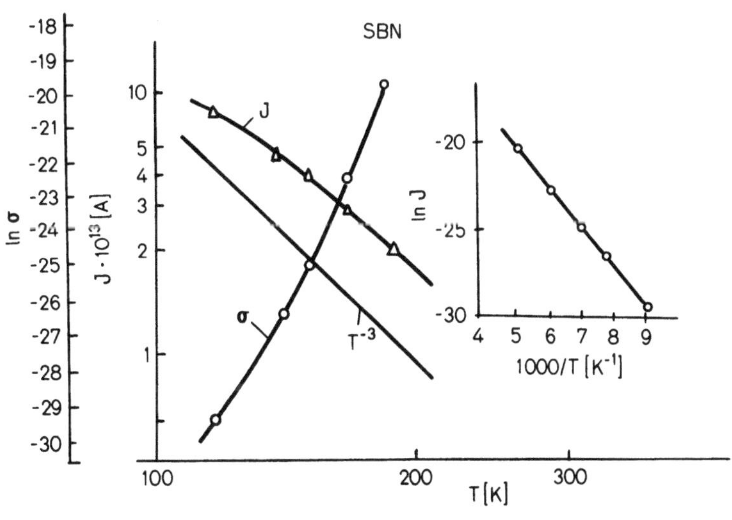

Fig. 6.12. Temperature dependences of J and σ_{ph} for SBN

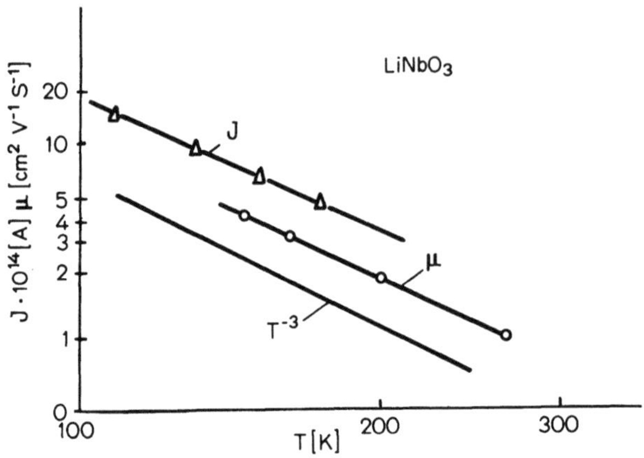

Fig. 6.13. Temperature dependences of J and μ for $LiNbO_3$

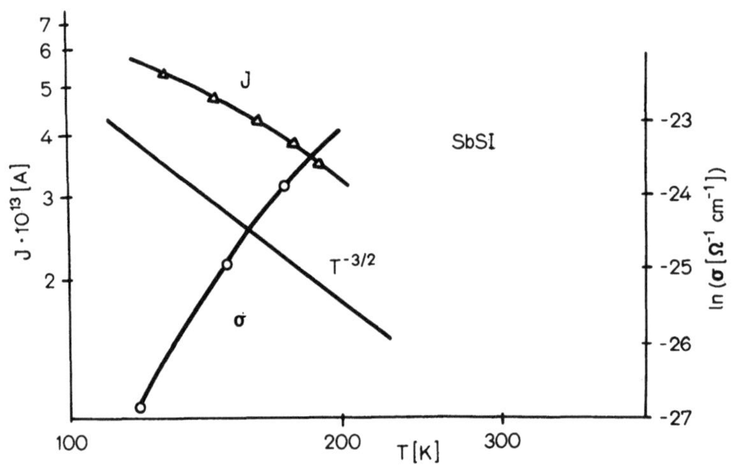

Fig. 6.14. Temperature dependences of J and σ_{ph} for SbSI

Figure 6.16 shows the spectral distribution of the photovoltaic current J, the photocurrent J_{ph}, and the induced field $\tilde{\mathscr{E}}$ in $LiNbO_3$:Fe in the oxidized and reduced state [6.26]. As in [6.24], the spectrum of J shows a maximum at $\lambda \simeq 400$ nm, corresponding to electron transition from the Fe^{2+} levels to the conduction band, which is not observed in the spectrum of J_{ph}.

Investigation of the EPR spectra [6.7] and the Mössbauer effect [6.27] in $LiNbO_3$:Fe indicated the presence of a complex Fe^{2+} - O vacancy oriented in the [001] direction; here, Fe^{2+} substitutes for Nb^{5+}. Calculation of the energy of this center yielded a value of 3.2 eV [6.28], which agrees with the spectra of J for $LiNbO_3$:Fe given in Figs. 6.7, 16 and in Fig. 6.17. Much

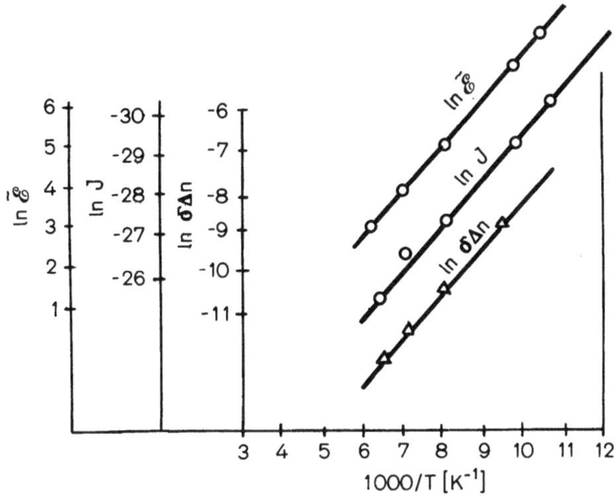

Fig. 6.15. Temperature dependenced of $\ell n\tilde{\mathscr{E}}$ (1), $\ell n\sigma$ (2) and $\ell n\Delta n$ (3) for $Ba_{0.25}Sr_{0.75}Nb_2O_6$

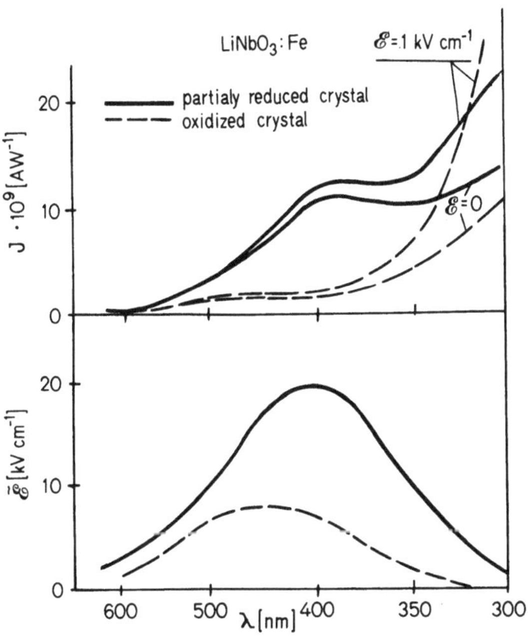

Fig. 6.16. Spectral distribution of photovoltaic current ($\mathscr{E}$ = 0), photocurrent in [001] direction in external field $\mathscr{E}$, and of photoinduced field $\tilde{\mathscr{E}}$ for $LiNbO_3$:Fe [6.26]

of the available data point to the donor and acceptor nature of Fe^{2+} and Fe^{3+}, respectively [6.29].

According to [6.25,26], as distinguished from [3.21], J increases outside the intrinsic absorption edge ($\lambda \simeq$ 320 nm) toward the shorter wavelengths, although more slowly than J_{ph}. In conformity with (6.4) this results in a broad maximum in the spectrum of the field $\tilde{\mathscr{E}}$. These data indicate a substantial role of band-band transitions and show that consequently the

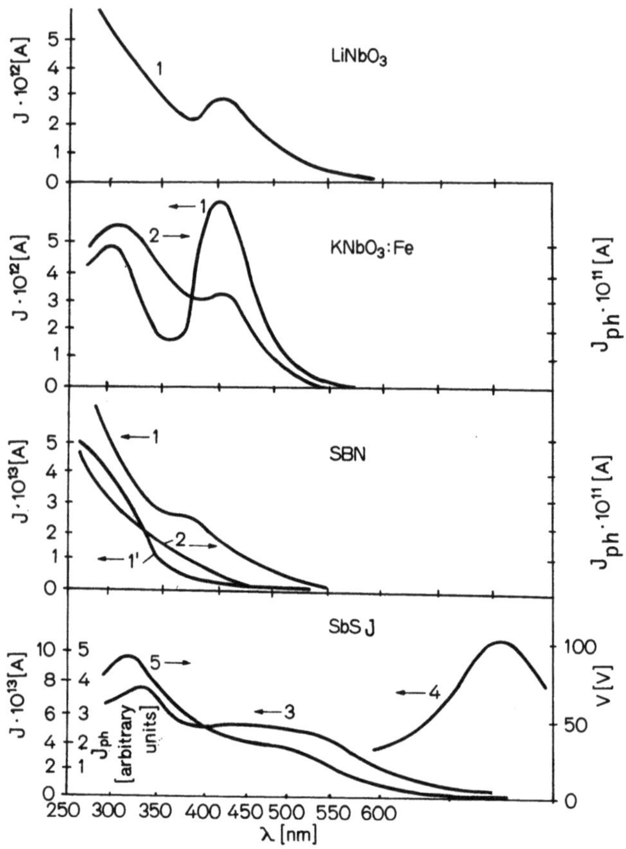

Fig. 6.17. Spectral distribution of photovoltaic current J and photocurrent J_{ph} [001] direction in $LiNbO_3$: Fe, $KNbO_3$: Fe (0.1%), SBN and SbSI. (1) J; (2) J_{ph} at T = 300 K; (1') J at T = 103 K; (3,5) J and V, respectively, for SBSI at T = 113 K; (4) J_{ph} at 293 K

AP effect in lithium niobate may be either of an impurity of an intrinsic nature.

This conclusion is supported by the spectra of J and J_{ph} obtained for $LiNbO_3$:Fe, $KNbO_3$:Fe, SNB, and SbSI (Fig.6.17) [6.23]. For instance, the spectrum of J and J_{ph} for $KNbO_3$:Fe shows both an impurity maximum Fe^{2+} - O vacancy at λ = 400 nm [6.7] and an intrinsic maximum. A comparison of the spectra of J and J_{ph} for the ferroelectric SbSI reveals a substantial difference in the behavior of these spectra near the intrinsic absorption edge ($E_g \simeq 2$ eV). While J_{ph} forms a sharp maximum due to a decrease in lifetime τ because of the surface recombination [1.14], the photovoltaic current slightly increases toward the shorter wavelengths. According to (6.4) this leads to an abrupt increase of the photovoltage in SbSI in the region of strong absorption.

In [3.21] a photovoltaic element based on a homogeneous pyroelectric crystal was considered as an energy converter. The efficiency of such an unloaded converter per unit volume of the crystal is given by the relation

$$\eta = \frac{J^2}{\alpha I \sigma} \tag{6.9}$$

where α is the absorption coefficient. For $LiNbO_3$:Fe η = 0.03%; so lithium niobate cannot be considered as a useful material for energy conversion.

The AP effect is evidently common to all pyroelectric crystals. According to (6.4), however, anomalously high photovoltages are observed only with a sufficiently low photoconductivity. When the photoconductivity is high, the AP effect can be detected only by measuring the temperature and intensity-current characteristics of the photovoltaic current. This possibly explains why the AP effect has not been observed previously in such pyroelectrics as CdS or ZnS in the hexagonal phase. The APV effect observed early on films and crystals of ZnS, is associated with the stratified structure and the composition of the elementary photo e.m.f. appearing at the boundary of the cubic and hexagonal phase [6.30,31]. For the same reason we do not consider here the APV effect in ferroelectric ceramics [6.32].

6.3 The Nature of the AP Effect in Ferroelectrics

6.3.1 Asymmetry of Impurity Centers and Frank-Condon Relaxation

We shall now discuss a model for the anomalous photovoltaic effect based on electron transitions to the band from an impurity center with an asymmetric potential distribution [3.21,6.2]. This leads to an asymmetric ejection of an electron from the impurity level to the band, and accordingly to a stationary photovoltaic current.

Suppose the potential of the center is in the form of an asymmetric rectangular barrier shown in Fig. 6.18 [3.21]. Excitation of an electron between the states ξ_0 and ξ_1 (for $E < V_2$) leads to a shift of the localized electron, as shown with arrows in Fig. 6.18. This, in turn, may result in a change of the spontaneous polarization (Sect.3.5). If the excited electron has an energy of $V_2 < E < V_1$, it shifts into a free state with a wave vector $+\underline{k}$ (the direction of the wave vector is parallel to the spontaneous polarization). An electron with a wave vector $-\underline{k}$ only partly penetrates through the potential barrier. Thus, the probability p+ of motion of the electron in

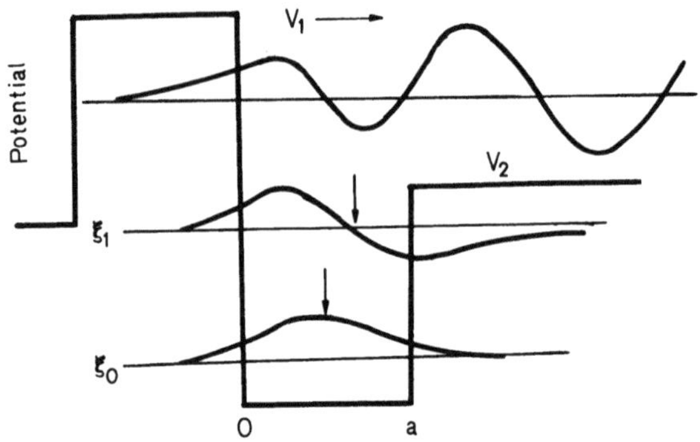

Fig. 6.18. Asymmetric potential of impurity center in ferroelectric and corresponding energy and wave functions [3.21]

the direction $+\underline{k}$ differs from the probability p- of its motion in the direction $\underline{k}$, and the more so, the larger is the spontaneous polarization.

After the electron is excited into the band, the ionized center shifts along the direction of the spontaneous polarization because of the interaction with the soft optical mode (Frank-Condon relaxation). This mechanism gives rise to the additional component of the photovoltaic current.

Since the asymmetry of the potential of the impurity center is determined by the direction of the spontaneous polarization and is the same for all the centers, for exciting light of intensity I and frequency ω_0, the density of the photovoltaic current J_1 associated with the asymmetric ejection of the electron is

$$J_1 = \frac{q\alpha I}{\hbar\omega_0}(p_+\ell_+ - p_-\ell_-) \quad , \tag{6.10}$$

where $\ell_\pm$ is the free path length of the electron in the direction $\pm\underline{k}$, and α is the light absorption coefficient. It is easy to relate J_1 with the spontaneous polarization P_0. Relating the probabilities p_+ and p_- with the tunnel penetration through rectangular barriers of height V_1 and V_2 and assuming $V_2 \ll V_1 \sim P_0^2$, we have

$$J_1 \simeq \frac{q\alpha I}{\hbar\omega_0}\langle\ell\rangle P_0 \quad , \tag{6.11}$$

where $\langle\ell\rangle$ is the average shift of the electron. The density of the photovoltaic current J_2, associated with the motion of the ionized impurity

(Frank-Condon relaxation), is

$$J_2 = \frac{\alpha I}{\hbar\omega_{0i}} \sum_i Z_i \Delta\ell_i \tag{6.12}$$

where Z_i is the charge of the ith ion, and $\Delta\ell_i$ is the displacement of the ith ion, which is proportional to P_0. The summation in (6.12) is over all the ions.

The currents associated with the recombination of the photoexcited carriers must be subtracted from the currents (6.10) and (6.12). Indeed, after the scattering of the photoexcited carrier, its directed momentum $\underline{k} = 0$, and it does not make any contribution to the photovoltaic current before recombination. Denoting the probability of recombination in the direction $+\underline{k}$ and $-\underline{k}$ by p'_+ and p'_-, respectively, we obtain the corresponding recombination currents from (6.10) and (6.12) by replacing p_+ and p_- with p'_+ and p'_- and Z_i with Z'_i. It is implied here that $\Delta\ell'_i = \Delta\ell_i$ and $Z'_i \neq Z_i$, since the ion returns to the initial state and has a different charge before and after recombination. The final expression for the photovoltaic current J is [3.21]

$$J = J_1 - J_2 = k_1 \alpha I \quad , \tag{6.13}$$

where the Glass constant

$$k_1 = \frac{q}{\hbar\omega_0} (\ell_+ p_+ - \ell_- p_- + \ell'_+ p'_+ - \ell'_- p'_-) + \sum_i \frac{Z_i - Z'_i}{\hbar\omega_0} \Delta\ell_i \tag{6.14}$$

depends exclusively on the nature of the impurity center, the free path length of the electron, and the photon energy.

It is significant that even at $V_1 = V_2$, i.e. when there is no asymmetric ejection of an electron and $J_1 = 0$, the photovoltaic current $J = J_2$ is nonzero because of the interaction of the charged center with the soft optical mode.

The current-voltage characteristics of $LiNbO_3$:Fe there corresponds in Fig. 6.8 to $k_1 = 2.5 \times 10^{-9}$ A cm.W^{-1} and $\sigma_d + \sigma_{ph} = 1.3 \times 10^{-14} + 1.2 \times 10^{-12}$ I Ω^{-1} cm^{-1} [3.21]. This value is in good agreement with the photovoltages calculated for $LiNbO_3$:Fe by (6.4). According to the same data the Glass constant k_1 for $LiNbO_3$:Fe is independent of the concentration of the donor centers of Fe^{2+} and of the ratio between the concentrations of Fe^{2+} and Fe^{3+}. At the same time this ratio strongly affects the electrical and photoconductivity of the crystal and, hence, the APV effect. When $LiNbO_3$:Fe is reduced in a nitrogen atmosphere and the concentration of Fe^{2+} due to the concentration of Fe^{3+} is increased, values of σ_d and σ_{ph} increase, while those of V and $\tilde{\mathscr{E}}$ decrease correspondingly.

Substituting into (6.10) $\ell_{\pm} = (\tau_0 v)_{\pm}$, where $\tau_0 = (m/q)\mu$ is the relaxation time, v is the electron velocity, and m is the effect mass, we have

$$J_1 = \frac{q\alpha I}{\hbar\omega}\Delta(p\ell) = \frac{\alpha Im}{\hbar\omega}\Delta(pv\mu) \quad . \tag{6.15}$$

Under the same conditions the photoconductivity σ_{ph} has the form

$$\sigma_{ph} = \frac{q\alpha I}{\hbar\omega}\tau\mu \quad , \tag{6.16}$$

where τ is the electron lifetime in the band. Thus, the photovoltaic current J_1, associated with the asymmetric ejection of the electron, is independent of the lifetime τ, which, in turn, is a function of the temperature and light intensity. This explains the difference in the characteristics of J and J_{ph} mentioned above. While the intensity-current characteristic of the photovoltaic current (6.14) is linear, the shape of the intensity-current characteristic of the photocurrent (6.16) depends on the lifetime $\tau = \tau(I)$, i.e. on the recombination mechanism [3.6]. The same mechanism determines the temperature dependence of the photocurrent. According to (6.15) the weak temperature dependence of the photovoltaic current must be due to the mobility μ, i.e. to scattering.

Substituting the expression for the photovoltaic current J (6.13) into (6.9), we get, for the energy conversion efficiency η:

$$\eta = \frac{m\langle\ell\rangle^2}{\tau\tau_0\hbar\omega_0} \tag{6.17}$$

The conversion efficiency is maximal when the lifetime of the excited carriers does not exceed the relaxation time. For the ferroelectrics investigated up to now $\tau \gg \tau_0$, which explains the low values of η.

The experimental values of k_1 in ferroelectric niobates correspond to an average electron shift of $\langle\ell\rangle \simeq 1$ Å [3.21,6.7]. It should be emphasized that with such small shifts the question of the applicability of the band model and approximation of the effective mass remains open, so in the Glass model the quantity $\langle \ell \rangle$ should be taken with the average shift in the jump mechanism (a small-radius polaron [1.7]). It is worth mentioning here that in deriving (6.14) it was implicitly assumed that the quantum yield of the electrons $\gamma = 1$. RUPPEL et al. [6.33] measured the quantum yield of the electrons to the conduction band from the Fe^{2+} centers in $LiNbO_3$ and showed that $\gamma \lesssim 4 \times 10^{-3}$. From (6.14) this leads to an average shift of $\langle \ell \rangle \gtrsim 40$ Å which greatly exceeds the average jump length and thus contradicts the Glass model. To explain the nature of the photovoltaic current, the model of photoinduced fluctuations, discussed later on, was used.

The photovoltaic current considered in the Glass model [3.21,6.2] is essentially an impurity current, since it is due to the asymmetry of the potential of the impurity center or to its interaction with the transverse optical vibration (Frank-Condon relaxation). In [6.34] the Frank-Condon relaxation mechanism was generalized for the case of band-band transitions, thereby indicating that the photovoltaic current can have an intrinsic nature. According to the model [6.34] an electron shifts from ground state to the excited Frank-Condon state under the influence of a quantum of the intrinsic light, $\hbar\omega_0$ (Fig. 6.19). On radiationless transition of an electron from the excited to the relaxation state it interacts with the soft optical mode, which results in a shift of the active ion by a distance of $\Delta\ell$. This induces a photo e.m.f. field $\mathscr{E}_{ef}$, which can be obtained directly from the relaxation energy.

$$\mathscr{E}_{ef} = \frac{1}{2}\,\frac{M\omega^2\Delta\ell}{z'} \tag{6.18}$$

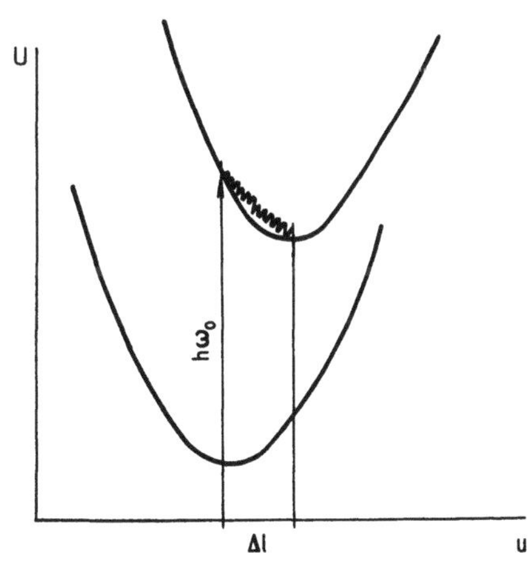

Fig. 6.19. Frank-Condon mechanism of photovoltaic effect [3.21]

where M and z' are the mass and charge of the active ion, respectively, and ω is the soft mode frequency. Therefore the expression for the photovoltaic current J_2 can be transformed to [6.34]

$$J_2 = \frac{\alpha I}{\hbar\omega_0} Z'\Delta\ell = 2\frac{\alpha I}{\hbar\omega_0}\frac{(Z')^2}{M\omega^2}\mathscr{E}_{ef} \quad . \tag{6.19}$$

It was shown in 6.35,36 that a photovoltaic current is induced in a photoferroelectric not only by asymmetry of the impurity centers responsible for the generation and recombination of nonequilibrium carriers, but also by asymmetry of scattering on impurities and phonons. Asymmetry of elementary electron processes in a photoferroelectric is, in turn, associated with the asymmetric shape of the potential of impurity centers and their identical orientation in the lattice with respect to the direction of the spontaneous polarization. The variation in the concentration and distribution of nonequilibrium electrons in the conduction band is described by the kinetic equation for the distribution function f_k:

$$\frac{\partial f_k}{\partial t} = I_k^e - I_k^r + I_k^i + I_k^{ph} \quad ,$$

where I_k^e and I_k^r are the electron excitation and recombination rates, respectively, and I_k^i and I_k^{ph} are the numbers of collisions of an electron with impurities and phonons, respectively, per unit time. If the distribution function is symmetric and satisfied the condition $f_k = f_{-k}$, there is no current in the crystal. If the right-hand side of the kinetic equation contains an asymmetric term satisfying the condition $I_k^{as} = -I_{-k}^{as}$, then the stationary solution of the equation also contains an asymmetric term, i.e. the asymmetric part of the distribution function $f_k^{as} = -f_{-k}^{as}$. The presence of the asymmetric distribution function results in a stationary current of

$$J = \frac{q}{\hbar}\int \frac{\partial \varepsilon_k}{\partial k} f_k^{as}\, dk \quad .$$

Suppose the impurity center possesses a dipole moment μ_0. Then its potential is asymmetric and has the form

$$\varphi = q\,\frac{\mu_0 \cdot r}{\varepsilon r^3} \quad .$$

Not only excitation and recombination, but also scattering on a dipole center with an asymmetric potential is asymmetric and, hence, leads to the appearance of an asymmetric part in the distribution function, i.e. to a stationary current [6.35,36]. It is important to recall that up to now we have always dealt with the nonequilibrium distribution function. In [6.36] it was shown rigorously that for equilibrium electrons $f_k^{as} = 0$ and, consequently, the equilibrium (dark) current is equal to zero. This corresponds to the fact that in the equilibrium case (6.12) the current associated with the excitation of electrons from asymmetric centers is fully offset by the current due to electron recombination and scattering. This theory thus leads both to an extrinsic and a intrinsic photovoltaic current. In the latter case (band-band transitions) the photovoltaic current is proportional to the square of the relaxation time $J \sim \tau_0^2$. Thus, the efficiency η at intrinsic absorption must be lower than for the impurity effect.

Despite the fact that the experimental data [6.23,26,33] point to the presence of an intrinsic photovoltaic current, further investigation is required. For instance, attention is drawn to the increase of the photoconductivity and the photovoltaic current in the ultraviolet region of the spectrum, which corresponds to strong surface absorption (Fig.6.17). This effect is evident in all investigated oxygen-octahedral ferroelectrics, and its origin is not yet clear.

6.3.2 Photoinduced Fluctuations

Excitation of nonequilibrium electrons results in a change of spontaneous polarization. This effect, which stems from the thermodynamics of a ferroelectric semiconductor (Sect.2.3), may arise through various microscopic mechanisms (Chap.3). In particular, it may be caused by a change in the dipole moment of the centers [3.14,6.37] because of the excitation of the electrons in the centers, which have the same potential everywhere in the crystal, whose asymmetry is associated with the spontaneous polarization. As an example, the change in the dipole moment of the center is shown by arrows in Fig. 6.18. Localization of an excited electron in a trap may result in a local change of the spontaneous polarization, i.e. local fluctuation. The mechanism of interaction of a localized electron with a transverse optical phonon (the Jahn-Teller effect), leading to the formation of photoinduced fluctuations near the trap, has been considered [6.38].

In [6.12,14] it was assumed that illumination of $BaTiO_3$ in the intrinsic region results in the formation of photoinduced fluctuations in the shape of ellipsoids of diameter ~ 20 Å and length 100 Å. It was shown previously that such equilibrium thermal fluctuations exist in $BaTiO_3$ [6.39], and their volume V was estimated from data on optical absorption [1.14]. The authors of [5.26] detected photoinduced Rayleigh scattering in c-single-domain crystals of $BaTiO_3$ and showed that scattered light is polarized in the direction of the spontaneous polarization. This enabled them to relate photoinduced scattering in $BaTiO_3$ with photoinduced fluctuations of polarization.

Because of the change in spontaneous polarization, ΔP_0, an electric field $\mathcal{E} = 4\pi\Delta P/\varepsilon$ is produced in the fluctuation volume V, which has the same direction for all fluctuations because of symmetry. The transport of nonequilibrium electrons may be accomplished by the fluctuation field, and this leads to a photovoltaic current J. Since the lifetime of the fluctuations, τ_f, is much less than the Maxwellian relaxation time $\tau_f << \tau_1$, the screening of the field in the fluctuation volume can be neglected. In this case

$$J = \frac{\Delta P}{\varepsilon} q\mu n\, VN^* \tag{6.20}$$

where N^* is the concentration of localized fluctuations and n is the concentration of free nonequilibrium electrons. If each electron that makes a contribution to J is associated solely with its own fluctuation, as assumed in [6.12,14], then

$$J = \frac{\Delta\mu_0}{V\varepsilon} q\mu N^* \tag{6.21}$$

where $\Delta\mu_0$ is the change of the dipole moment of the center near which the fluctuation is localized.

In this last model of photoinduced fluctuation, the excited electron is associated exclusively with the fluctuation induced by it, which leads to the expression (6.21) for the photovoltaic current. To see this, consider Fig. 6.18. Suppose the light absorption leads to a transition of the electron in the impurity center from the ground to the excited state. The corresponding change in the dipole moment of the center by $\Delta\mu_0$ (Sect.3.4) results in the formation, near the center, of a localized region of volume V_0, where the spontaneous polarization is changed by $\Delta P_0 = \Delta\mu_0/V_0$ and there is a field $\mathscr{E} \simeq 4\pi\Delta P_0/\varepsilon$ (photoinduced fluctuation). If the sign of ΔP_0 (this sign is identical for all impurity centers for reasons of symmetry) corresponds, according to (2.46), to the shift of the band-edge energy towards the excitation level, the excited electron finds itself in the band, where it makes a shift under the influence of the field $\tilde{\mathscr{E}}$ within the fluctuation range. Another possible conduction mechanism is the polaron mechanism, when the photoexcitation simultaneously induces fluctuation and transition of the excited electron to the polaron band [1.7].

Irrespective of the conduction mechanism, the kinetics of the formation of fluctuations are in this case described by the equation

$$\frac{dN^*}{dt} = \frac{sI}{\hbar\omega}(M - N^*) - \frac{N^*}{\tau_e}$$

where N^* is the concentration of the photoinduced fluctuations, M is the concentration of the impurity centers, τ_e is the fluctuation lifetime, which is equal to that of the electron in the excited state, I is the light intensity, and s is the cross section of interaction of the impurity center with the photon. The corresponding stationary concentration of fluctuations ($dN^*/dt = 0$) is given by

$$N^*_{st} = \frac{\frac{sI}{\hbar\omega}}{\frac{sI}{\hbar\omega} + \frac{1}{\tau_e}} M$$

Substituting N^*_{st} into (6.20) yields the expression for the photovoltaic current. It is easy to see that for $1/\tau_e \simeq 10^4 - 10^6\ s^{-1}$ (see Sects.5.3,5;7.1), $s \simeq 10^{-15}\ cm^2$ [1.14], and $I \leq 1\ W.cm^{-2}$, the photovoltaic current depends linearly on I, since $sI/\hbar\omega \ll 1/\tau_e$:

$$N^*_{st} \simeq \frac{sI}{\hbar\omega}\tau_e M$$

$$J \simeq \frac{\Delta\mu_0}{V\varepsilon} q\mu \frac{sI}{\hbar\omega}\tau_e M \quad . \tag{6.22}$$

Substituting into (6.22) the values of the constants for $LiNbO_3:Cr^{3+}$ ($\Delta\mu_0 \simeq 10^{-27}$ C cm, $M \simeq 8 \times 10^{19}\ cm^{-3}$, $\tau_e \simeq 10^{-5}$ s [5.18,6.6], $s \simeq 10^{-15}\ cm^2$

[1.14], $\varepsilon \simeq 30$, $\hbar\omega \simeq 5 \times 10^{-12}$ erg), (6.21) reduces to

$$J[\mathrm{Acm}^{-2}] = \frac{\mu}{V} I[\mathrm{Wcm}^{-2}] \qquad (6.23)$$

Let us compare (6.23) with the experimental intensity-current characteristic of the photovoltaic current for $LiNbO_3$:Fe (6.8). To do this, we substitute into (6.23) the value of mobility for $LiNbO_3$, for which the literature gives contradictory data. Thus, substituting $\mu \simeq 10^{-3}$ cm^2 V^{-1} s^{-1} [6.33], we obtain coincidence of (6.23) with (6.8) for $V \simeq 10^{-24}$ cm^3, which corresponds to an average fluctuation radius of $\langle \ell \rangle \simeq 1$ Å. So, here $\langle \ell \rangle$ is close to the average shift of the electron in the asymmetric center model. If we use $\mu \simeq 0.8$ cm^2 V^{-1} s^{-1} [6.25], (6.22) coincides with (6.8) at $V \simeq 10^{-21}$ cm^3, or $\langle \ell \rangle \simeq 10$ Å. Finally, if we take $\mu \simeq 15$ cm^2 V^{-1} s^{-1} [6.40] then $V \simeq 10^{-20}$ cm^3 and $\langle \ell \rangle \simeq 10 - 100$ Å, which corresponds to the assumption about the dimension of photoinduced fluctuations in $BaTiO_3$ [6.12,14].

The spread in the values of μ for $LiNbO_3$ may be due to the different concentration of the impurity centers and other defects and to the different conditions of reduction of crystals in a hydrogen atmosphere. It is also important to recall that in comparing (6.22) with (6.8) we used some parameters corresponding to excitation of Cr^{3+} ions in $LiNbO_3$, while the mobility was that of $LiNbO_3$:Fe. As for $BaTiO_3$:Fe, according to the data on photoinduced Rayleigh scattering, $\tau_e \simeq 10^{-4} - 10^{-5}$ s near the Curie point (Sect.5.5), if we associate the photoinduced fluctuations in $BaTiO_3$:Fe with the model proposed here. Unfortunately, owing to the lack of data on the variation $\Delta\mu_0$ in the dipole moment of the iron centers in $BaTiO_3$:Fe, we cannot estimate the dimension of photoinduced fluctuations in barium titanate with respect to the photovoltaic current. According to (6.22) the temperature dependence of J is determined by that of the mobility μ and τ_e. Both dependences lead to an increase in J with decreasing temperature. The equilibrium current for the fluctuation model is nonexistent for the same reason as for the asymmetric center model.

We have seen that the APV effect, at least for the ferroelectrics investigated so far, is associated with the presence of a stationary photovoltaic current in the ferroelectric. This current can be explained only by introducing a radically new carrier transport mechanism. Some of these mechanisms were discussed above.

In conclusion, we point out the existence in a ferroelectric of an APV effect not associated with the photovoltaic current and having the same nature as other photoferroelectric phenomena. As noted in Chap. 5, photoferroelectric phenomena may be based on either one of two principal mechanisms: first, the screening of the spontaneous polarization by nonequilibrium carriers and the resulting formation of space charges and internal fields; second, the direct influence of nonequilibrium carriers on the spontaneous polarization, for instance, due to the Jahn-Teller effect or to interaction of recharged centers with the lattice. It is significant, however, that these mechanisms admit of the existence of a transitional photocurrent, as distinct from a stationary one.

7. The Photorefractive Effect in Ferroelectrics

From the phenomenological (Sect.2.3) and microscopic (Chap.3) discussion of the influence of nonequilibrium carriers on the ferroelectric properties of crystals, we deduce that illumination must influence the spontaneous polarization P_0 and, in turn, lead to the corresponding change in birefringence (the Pockels effect). The formation of strong fields $\mathscr{E}$ in a high-resistance ferroelectric as a result of an anomalous photovoltaic effect (AP effect) may also cause a change in birefringence. Therefore the influence of intense illumination on birefringence in ferroelectris, which was discovered relatively recently [7.1,2], can be considered as a genuine photoferroelectric phenomenon.

As noted above, the effect of a photoinduced change in birefringence is referred to as optical damage or the photorefractive effect (PR effect) in the literature and has been widely used for recording volume holograms. In practice, the PR effect consists in the following. As a result of local illumination of a ferroelectric crystal with an intense transient light (focused laser beam) a reversible change in birefringence occurs in the crystal volume inside the light beam, mainly because of the change in the refractive index n_e. The extent of this change may be as great as 10^{-4} - 10^{-3} for some pyroelectrics ($LiNbO_3$, $LiTaO_3$), and the storage time of the effect varies over a very wide range, from milliseconds for $BaTiO_3$ to months for $LiNbO_3$. Hologram recording is accomplished by means of volume modulation Δn corresponding to that of the recording beams the recording resolution is extremely high, 10^2 - 10^4 lines/mm.

To observe the PR effect, some general conditions must be fulfilled. First of all, for the value of Δn to be detectable, the electro-optic coefficients of the crystals under investigation must be high enough. Since the effect is a volume effect, the photoexcitation wavelength must be far enough from the intrinsic absorption edge, which does not exclude the possibility of obtaining the effect under illumination with intrinsic light. Therefore, in investigating the PR effect, which is, as a rule, observed in materials with very broad bands, He-Ne (0.628 μm) or Ar (0.488 μm) lasers are usually employed as light sources. The appearance of the PR effect does not require coherence of the exciting light; the main condition for an appreciable change in birefringence is sufficiently high radiation energy. The value of Δn also depends on the exposure time. In $LiNbO_3$, at small exposures of ~ 0.01 s, the value of Δn is proportional to the product of the exposure time by the radiation intensity, $\Delta n \sim It$, up to very high intensities [7.3]. At prolonged exposures of ~ 1 min we have $\Delta n \sim I^{1/2}$, and saturation is achieved at intensities of 20 mW (Fig. 7.1). In investigations of the PR effect, long exposure and low intensity are usually used.

The PR effect may either have an intrinsic or an impurity origin. The nature of the levels responsible for an impurity PR effect has been investigated mainly for $LiNbO_3$ [7.4,5]. Like $BaTiO_3$, lithium niobate is an oxygen-octahedral ferroelectric, in which the valence band is formed by the overlap of oxygen orbitals, and the conduction band, by that of niobium ion orbitals. As indicated in Chap. 6, investigations into the absorption spectra of pure and iron-alloyed $LiNbO_3$ [7.5] revealed that $LiNbO_3$:Fe has localized Fe^{2+} and Fe^{3+} levels with Fe^{2+} being a donor, while Fe^{3+} is a trap. The PR effect in $LiNbO_3$:Fe is attributed to the photoexcitation of electrons from the Fe^{2+} levels into the conduction band, after which the electrons are captured by Fe^{3+} traps. Thus, the photorefractive sensitivity of $LiNbO_3$ increases with the concentration of Fe^{3+} traps and is independent of the concentration of Fe^{2+} donors [6.2,7.5,6]. The thermal activation energy of annealing at

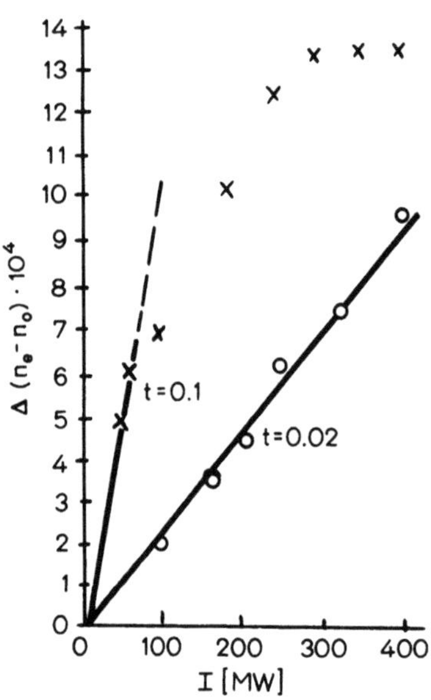

Fig. 7.1. Dependence of $\Delta(n_e - n_0)$ in lithium niobate on light intensity for exposures of 0.02 s and 0.1 s. Laser beam diameter is 10^{-2} cm

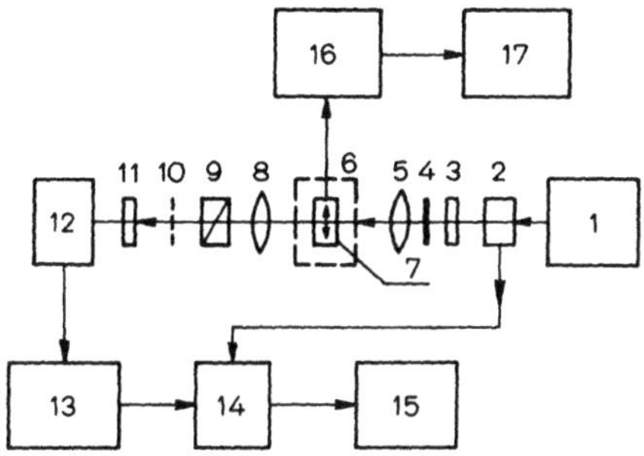

Fig. 7.2. Block diagram of set-up for investigating photorefractive effect and short-circuit photocurrents. (1) He-Ne laser; (2) Laser beam modulator; (3) Neutral filter; (4) Polarizer; (5) Lens; (6) Cryostat; (7) Crystal; (8) Lens; (9) Quartz wedge; (10) Analyzer; (11) Neutral filter; (12) Photoelectric multiplier; (13) Alternating-current amplifier; (14) Synchronous detector; (15) Recorder; (16) Electrometrical amplifier; (17) Two-coordinate recorder

$T \simeq 170^{\circ}C$ for $LiNbO_3$ is 1.1 eV [7.4], but it has not yet been possible to relate this value with the activation energy of the Fe^{3+} capture levels.

There are a number of precision techniques that make it possible to measure Δn to within an accuracy of 10^{-6} [7.3]. For example, Fig. 7.2 shows a block diagram for measuring Δn by the compensation method. Another equivalent method consists in measuring the diffraction efficiency when recording an elementary phase hologram (see below).

The application of the PR effect in phase holography [7.1-4,7-14] is briefly discussed below. Here, we shall survey the investigations on the photoferroelectric nature of this phenomenon, following chiefly [6.6].

7.1 Mechanisms of the PR Effect in Ferroelectrics

A perceptible photoinduced change in birefringence in most ferroelectrics (KTN [7.3], $BaTiO_3$ [7.7], and PLZT [7.14]) is observed only when an external constant field of the order of several kilovolts per centimeter is applied to the crystal, and only in a limited number of cases (for instance, in $LiNbO_3$ and other niobates, $LiTaO_3$ [7.1,2], KDP [7.15]) the effect takes place in the absence of an external field. In the presence of an external field, the PR effect is due to the screening of the external field by nonequilibrium carriers, the redistribution of the field in the crystal volume, and the change in birefringence because of a linear or quadratic electro-optic effect. This mechanism is observed not only in ferroelectrics, for example, in the paraelectric phase of KTN [7.3] or PLZT [7.14], but also in nonferroelectric electro-optic materials, such as $Bi_{12} SiO_{20}$ [7.15], CdS [7.14]. In the absence of an external field, the very existence of the optical damage effect is due to the ferroelectric nature of the crystal. We shall consider this particular case first.

Like all photoferroelectric phenomena, the optical damage effect may be associated with one of three principal independent mechanisms. These are: first, the anomalous photovoltaic effect; second, optical recharging of the centers, which causes, among the other photoferroelectric phenomena, a change in spontaneous polarization [see (2.37)]; third, the screening of spontaneous polarization, which is accompanied by spatial separation of the carriers and the formation of internal fields of space charges. Investigations into the microscopic nature of the PR effects are, in general, focused on these three possible mechanisms. We do not mention the diffusion mechanism here [7.17], because it may also take place in nonferroelectric electro-optic photoconductors, nor do we discuss the ordinary pyroeffect. These mechanisms will be briefly considered below.

The first attempt at finding a quantitative relationship between the PR effect and the photoconductivity of ferroelectrics was made on the basis of the screening mechanism for $LiTaO_3$ and $LiNbO_3$ [7.3]. It was assumed that on local illumination of a crystal, carriers photoexcited from some levels drift, under the action of an internal field, [which is always present in the ferroelectric volume, irrespective of the screening conditions (Sect. 4.3)], to the periphery of the area of illumination, where they are captured by deep levels. The spatial separation of the carriers, due to the screening of the internal field, leads to the formation of a local field $\tilde{\mathcal{E}}$ and, hence, to a local change in birefringence due to the linear electro-optic effect given by

$$\Delta n = \frac{1}{2}\,(n_0^3\, r_{13} - n_e^3\, r_{33})\,\tilde{\mathcal{E}} \sim N \quad , \tag{6.1}$$

where n_0, n_e are, respectively, the refractive indices of the ordinary and extraordinary beams, N is the concentration of the captured carriers, and r_{ij} are the electro-optic coefficients. At first glance, the experimental data agree with this qualitative model (Fig.7.3). In the first place, the effect is anisotropic and is substantial only with the illumination is perpendicular to the ferroelectric axis of the crystal, c, i.e. when the direction of the drift of the screening carriers is determined by that of the internal field. The spatial distribution of Δn agrees with the distribution of the local screening field (Fig.7.4). The calculation by the electro-optic equation (7.1) showed that in order to achieve the maximum values observed in

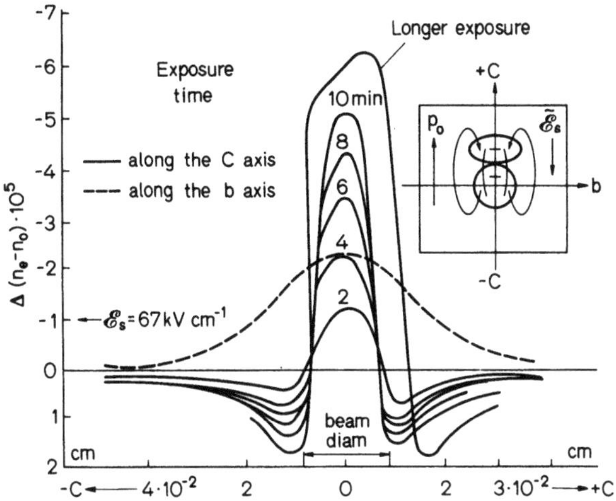

Fig. 7.3. PR effect in $LiTaO_3$. The inset shows the distribution of space charge $\tilde{\mathscr{E}}$ causing PR effect [7.3]

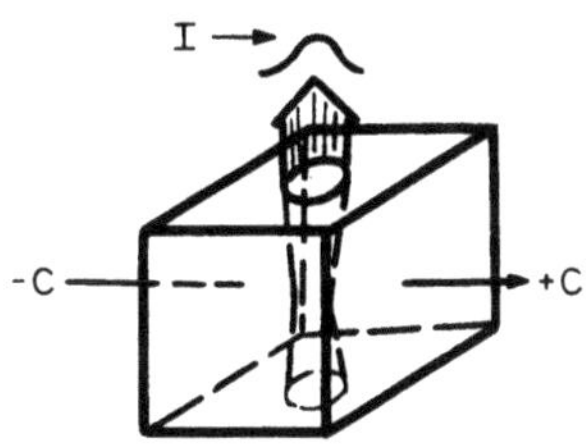

Fig. 7.4. Geometry of crystal and light beam

experiment $\Delta n \simeq 10^{-3}$, it is necessary that the crystal have an internal field $\tilde{\mathscr{E}} \simeq 7 \times 10^4$ V/cm [7.3]. The existence of such high macroscopic fields in a short-circuited ferroelectric seems doubtful. Besides, as noted in Chap. 6, the formation of optical damage in $LiNbO_3$ is accompanied by a stationary photovoltaic current, and not by a transitional screening current. Thus, as we shall see below, the AP effect, coupled with other possible mechanisms, explains all the photorefractive properties of $LiNbO_3$ without assuming the existence of an internal macroscopic field.

Let us consider the other mechanisms of the PR effect. Illumination of an n-type ferroelectric in an intrinsic or an impurity band leads to a photocurrent in the direction of spontaneous polarization (the z axis)

$$J = \sigma\mathscr{E} + k_1\alpha I + kT\mu_e \frac{\partial n}{\partial z} + \frac{\partial D_0}{\partial t} \quad . \tag{7.2}$$

Here, $\sigma = \sigma_d + \sigma_{ph}$ is the photoconductivity, μ_e is the electron mobility, n is the electron concentration in the band, k_1 is the Glass constant, α is the absorption coefficient, I is the light intensity, and $\mathscr{E}$ is the macroscopic field. The expression (7.2) differs from (6.3) by the diffusion current $kT\mu_e \partial n/\partial z$ and the displacement current $\partial D_0/\partial t$ due to the photoinduced change in the spontaneous displacement D_0. A macroscopic field $\tilde{\mathscr{E}}$

$$\tilde{\mathscr{E}} = \frac{4\pi}{\varepsilon} \int_0^t J dT \qquad (7.3)$$

is induced in the crystal within the exposure time t. Because of the linear electro-optic effect the field $\tilde{\mathscr{E}}$ leads to a PR effect

$$\Delta n_i = \frac{1}{2} n_i^3 r_{ij} \tilde{\mathscr{E}} \quad . \qquad (7.4)$$

The electro-optic equation (7.4) is recorded in the main crystallographic axes and is similar to (7.1). Irrespective of the specific mechanism of the PR effect, the value of Δn_i is directly proportional to the electro-optic coefficients r_{ij}. But the parameter $r_{ij}/\varepsilon \sim P_0$ does not vary over an excessively wide range for ferroeelectrics. Thus the value of the PR effect, or the photorefractive sensitivity dn_i/dw (w is the energy absorbed per unit volume of the crystal), is primarily determined by the photocarrier transport mechanism. We shall therefore consider successively the transport mechanisms associated with the AP effect, the influence of electron excitations on spontaneous polarization (displacement), the diffusion of non-equilibrium carriers, and the photoconduction in the external field.

7.1.1 The AP Effect

With an AP effect in $LiNbO_3$, KDP, and other high-resistance ferroelectrics, the field $\tilde{\mathscr{E}}$ may reach values of $\sim 10^5$ V/cm [6.2]. As shown in Chap. 6, the values of the fields $\tilde{\mathscr{E}}$ are in good agreement with the saturation values of Δn_i and the electro-optic coefficients. The temperature dependence of Δn_i correlates with that of the field $\tilde{\mathscr{E}}$ (for instance, for SBN and KDP). The kinetics of the PR effect (including its dark relaxation) and the dependence of Δn_i on the light intensity I are determined by the Maxwellian relaxation and the dependence $\tilde{\mathscr{E}} = \tilde{\mathscr{E}}(I)$. At high intensities I, when $\sigma_{ph} \gg \sigma_d$, $\tilde{\mathscr{E}}$ is independent of the light intensity, which corresponds to the saturation of the PR effect, when Δn_i no longer varies with I (Fig.7.1). Finally, the spectral distributions of the two effects coincide.

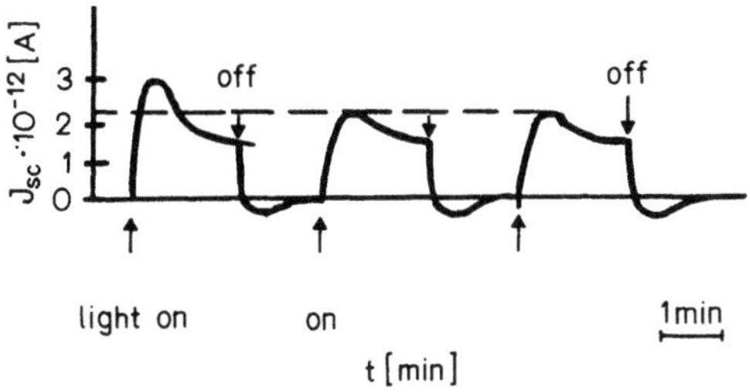

Fig. 7.5. Change in J_{sc} as a result of successive illuminations of crystal with laser beam at one point, in crystals of $Ba_{0.25}Sr_{0.75}Nb_2O_6$, T = 120 K

The relationship between the PR and AP effects for $LiNbO_3$ and SBN is also corroborated by simultaneous measurement of Δn_i and the transient photovoltaic current [7.18]. On switching on the light the photovoltaic current shows a transient maximum (Fig.7.5). On repeated switching on and off this maximum changes because of the formation of the space-charge field $\tilde{\mathscr{E}}$ (Fig. 7.5). By measuring this change one can estimate the value of the space charge $\Delta Q = \int J dt$ and of the field $\tilde{\mathscr{E}} = 4\pi\Delta Q/\varepsilon$ by integrating the transient photovoltaic current. We can substitute the field $\tilde{\mathscr{E}}$ thus calculated into the electro-optic equation (7.4) and estimate the PR effect Δn_i. Figure 7.5 depicts the change in the transient maximum of the photovoltaic current on successive illuminations of SBN with the aid of a He-Ne laser. Similar measurements were made in $LiNbO_3$ [7.18]. Table 7.1 lists the values of ΔQ and $\tilde{\mathscr{E}}$ obtained by this method and the calculated and measured values of Δn_i. The values $n_0 \simeq 2.31$, $n_e \simeq 2.30$, $r_{33} \simeq 13.4 \times 10^{-8}$ cm·V^{-1}, and $r_{13} \simeq 6.7 \times 10^{-10}$ cm·V^{-1} for SBN were used in calculating Δn_i [7.19] (similar constants for $LiNbO_3$ are given in Table 6.1).

Although the AP effect leads to the formation of high fields $\tilde{\mathscr{E}}$ and correspondingly large values of Δn_i (provided the photoconductivity σ is not too high), the mechanism of carrier transport in the photovoltaic effect is inefficient, and therefore the photorefractive sensitivity dn_i/dw is not high (see below). In $LiNbO_3$:Fe the value of the asymmetric shift of the electron per absorbed photon at λ = 514 nm is 0.5 Å, which corresponds to a change of ~ 2.4 Debye units in the dipole moment [3.21]. RUPPEL et al. [6.33], however,

Table 7.1. The space-charge field $\tilde{\mathscr{E}}$ in $LiNbO_3$ and SBN

Crystal	T[K]	ΔQ[μC]	$\tilde{\mathscr{E}}$[Vcm^{-1}]	Δn_i calculated	Δn_i experimental
$LiNbO_3$	300	4×10^{-6}	1800	2.1×10^{-5}	5×10^{-5}
SBN	110	2×10^{-5}	600	4×10^{-4}	1.1×10^{-4}

showed that the average shift of the electron in $LiNbO_3$ is much larger, if we assume that the quantum yield is much less than unity. This, in turn, makes us discard the model of asymmetric impurity centers in favor of the photoinduced fluctuations model (Sect.6.4). This problem requires further investigation.

7.1.2 The Influence of Electron Excitations on Spontaneous Polarization

As was shown in Chap. 3, the influence of electron excitations on spontaneous polarization may be based on different micromechanisms, including the change in the dipole moments of the impurity centers when they are excited or recharged. Thus it is suggested in [6.37] that the change in the dipole moments of the impurities, when opticaly recharged, leads to a change in the macroscopic polarization and a corresponding change in the birefringence inside the area of illumination, $\Delta n \sim \Delta P_0$. Inside this light spot a polarization field appears, due to the spontaneous polarization gradient (a field), causing a drift of nonequilibrium carriers beyond the light spot. At the boundary of the light spot, the carriers are captured by deep levels, and within the Maxwellian time τ_1 equilibrium is reached, when the field induced by the trapped carriers completely screens the depolarization field.

After the light is switched off, within a short time the excited electrons (localized in shallow traps or on excited levels of the impurity centers) recombine with ionized centers, as a result of which $\Delta P_0 \to 0$ inside the light spot. The screening charge at the periphery of the light spot, being formed by carriers trapped into deep levels, remains for a long time, the exact duration depending on the energy of these levels and the temperature. Thus, after the light is switched off, an electric field appears inside the light spot, changing the refractive index n_i by the electro-optic effect. This field $\sim (4\pi/\varepsilon)\Delta P_0$, where ΔP_0 is the initial change in spontaneous polarization inside the light spot. At equilibrium the stationary value of Δn_i is determined by the value of ΔP_0. Thus, in this model the two main mechanisms responsible for photoferroelectric phenomena (optical recharing of the levels and screening of spontaneous polarization) operate jointly. Some quantitative estimates were made in [6.37]. Using the maximal value of $\Delta n_i \sim 10^{-3}$ for $LiNbO_3$ found in [6.16], the author of [6.37] obtained from the electro-optic equations the value of the change in spontaneous polarization corresponding to this change in birefringence, $\Delta P_0/P_0 = 3 \times 10^{-3}$ ($P_0 \simeq 70\ \mu C/cm^2$ for $LiNbO_3$). Such a change in spontaneous polarization is possible only when at least 1-2% of unit cells in the $LiNbO_3$ crystal contain optically recharged effects (assuming that the value of the dipole moment of the photoexcited impurity is about 5% of the value of the dipole moment of the unit cell of the perfect polar crystal).

Experimental data on the influence of nonequilibrium carriers on spontaneous polarization were obtained when investigating photoexcitation processes in doped polar crystals under illumination with very short light pulses. The processes of relaxation of radiationless excitation of Cu^{2+} and Cr^{3+} impurities in $LiNbO_3$ and $LiTaO_3$ were investigated [5.17,18,7.6,20-22]. These impurities in a polar lattice possess a finite dipole moment μ_0. On illumination of crystals of $LiNbO_3:Cr^{3+}$, $LiNbO_3:Cu^{2+}$, $LiTaO_3:Cr^{3+}$, $LiTaO_3:Cu^{2+}$ with nanosecond pulses from a Nd laser ($\lambda \simeq 1.06$ µm), pyrosignals with very short relaxation times τ_e (~ 2 µs for $LiNbO_3:Cr^{3+}$, $LiTaO_3:Cr^{3+}$ and 30 ps for $LiTaO_3:Cu^{2+}$) were observed. Such signals cannot be the usual pyroresponse, whose relaxation time is ~ 1 ms, and they cannot be attributed to direct phonon excitation, which is characterized by times of less than 1 ps. The authors assume that the appearance of such fast pyroresponses is associated with the excitation of the Cr^{3+}, Cu^{2+} impurities. Indeed, optical investigations showed that the lifetimes in the excited state are 1 µs for Cr^{3+} and 30 ps for Cu^{2+} at room temperature, which is close to the values obtained for τ_e. With a temperature decrease to 10-20 K, the lifetime of the excited states increases accordingly to 11 µs and 450 ps. The temperature dependence of τ_r in radiationless conditions is similar. The authors conclude that on illumination of the crystal with a light pulse, the impurity is excited and interacts with phonons, changing its dipole moment by $\Delta\mu_0$. It was established by independent measurements that for $LiNbO_3:Cr^{3+}$, $\Delta\mu_0 \simeq 0.6$-0.8 Debye units and for $LiTaO_3:Cu^{2+}$, $\Delta\mu_0 \simeq 3$ Debye units. With a decrease in temperature, $\Delta\mu_0$ increases. In turn, a change $\Delta\mu_0$ makes a contribution to the macroscopic polarization of the crystal. Thus, the relaxation time of a fast pyroresponse, τ_e, is the time of vibron (electron-phonon) relaxation. This effect is not observed in nonpolar crystals, where $\mu_0 = 0$. In pure crystals of $LiNbO_3$ and $LiTaO_3$ no fast pyroresponse appears, and only the usual pyroeffect, associated with phonon absorption, is observed. The most significant fact is that no fast pyroresponse is observed in $LiNbO_3:Fe$ against the background of the photovoltaic current and that, consequently, the PR effect in $LiNbO_3:Fe$ cannot be associated with this mechanism.

The detailed mechanism of the PR effect associated with the influence of electron excitations on spontaneous polarization may be different. Two cases are of interest here. In the first case [6.9] the formation of free nonequilibrium carriers leads to a change in spontaneous polarization ΔP_0 and subsequent screening of the depolarization field $\tilde{\mathcal{E}} = 4\pi\Delta P_0/\varepsilon$ through capturing nonequilibrium carriers by deep traps. After the light is switched off one can observe a stable PR effect due to the field $\tilde{\mathcal{E}}$ and the linear electro-optic effect. In this case the relaxation time of the PR effect is determined by the depth of the traps. In the second case the electron excitation is not associated with the formation of free carriers, and the relaxation time of the PR effect is determined by the lifetime of the excited states of the impurity centers. The authors of [6.6;7.23]. for instance, observed, by measuring the diffraction efficiency η_0, a PR effect in $LiNbO_3$:Cr on excitation of the $^4T_2$ states of a Cr^{3+} ion, when no free carriers are formed, and hence, the first three terms in (7.2) can be neglected. The change in birefringence took place within the duration of the exciting light pulse, while the hologram lifetime was determined by the relaxation of the excited state. The changein dipole moment, $\Delta\mu_0$, was several Debye units ($\sim 10^{-25}$ C.cm/ per absorbed photon), which corresponds to $\Delta\mu_0$ in the photovoltaic effect in $LiNbO_3$, but is much less that $\Delta\mu_0 = q\mu\tau\tilde{\mathcal{E}}$ on electron shift in the external field, i.e. during photoconductivity in KTN (see below). By exciting $LiNbO_3$:Cr with a concentration of Cr^{3+} ions of $\sim 8 \times 10^{19}$ cm^{-3} at 20 K by microsecond pulses in the region of λ = 620 nm elementary holograms were obtained, with a lifetime of $\sim$ 5 μs corresponding to the relaxation time of the excited states $^4T_2$. To prevent the formation of free nonequilibrium carriers due to the two-photon absorption, the intensity of the exciting light was kept below 10 kW/cm^2.

The PR effect associated with the term $\partial D_0/\partial t$ in (7.2) is anisotropic, i.e. depends on the shape of the light beam and its position relative to the direction of spontaneous displacement $\underline{D}_0$. This is most obvious when the light spot has the shape of a strip perpendicular of parallel to the direction of $\underline{D}_0$ or (in the case of an elementary hologram) of interference strips with a $\underline{K}$ vector parallel or perpendicular to $\underline{D}_0$ ($|\underline{K}|$ is the space frequency of the interference strips, see below). If $\underline{K} \| \underline{D}_0$, then at the strip boundaries or on the lattice planes div $\underline{D} \neq 0$, and the corresponding depolarization field in the crystal volume changes the birefringence Δn_i due to the electro-optic effect. If $\underline{K} \perp \underline{D}_0$, then div $\underline{D}$ = 0, the depolarization field is zero, and there is no PR effect. This anisotropy was discovered by measuring the diffraction efficiency in $LiNbO_3$:Cr on excitation of the $^4T_2$ states of the Cr^{3+} ion [6.6;7.10, 23].

The anisotropy of the PR effect itself, however, does not enable one to separate the photovoltaic mechanism from that of the photoinduced change in spontaneous polarization. It is seen, for instance, from (6.4) that if the dark conductivity cannot be neglected as compared with the photoconductivity ($\sigma_d \lesssim \sigma_{ph}$), then $\tilde{\mathscr{E}}(\underline{K}\|\underline{D}_0) > \tilde{\mathscr{E}}(\underline{K}\perp\underline{D}_0)$ and, hence, both mechanisms lead to the same anisotropy.

7.1.3 The Diffusion of Nonequilibrium Carriers

In the absence of an external field, an alternative mechanism for the PR effect is the diffusion of free nonequilibrium carriers. The value of the field $\Delta\tilde{\mathscr{E}}$ induced in this case can be determined from the condition

$$q\mu n\Delta\tilde{\mathscr{E}} + qD_e \frac{\partial n}{\partial z} = 0 \quad , \tag{7.5}$$

where n is the concentration of nonequilibrium electrons, and D_e is the electron diffusion coefficient. Assuming for the diffraction lattice $n = n_0 \cos(Kz - \varphi)$ (see below), we have from (7.5) [7.17]

$$\Delta\tilde{\mathscr{E}} = \frac{D_e}{\mu} K = \left(\frac{kT}{q}\right)K \quad , \tag{7.6}$$

where k is the Boltzmann constant. It can be seen from (7.6) that the field $\Delta\tilde{\mathscr{E}}$ and, accordingly, the PR effect are directly proportional to the space frequency K. It has already been indicated that if we assume $\Delta n_i \simeq 10^{-3}$ to be the saturation of the PR effect in $LiNbO_3$:Fe [6.16], then the saturation field is $\Delta\tilde{\mathscr{E}} \simeq 7 \times 10^4$ V/cm. According to (7.6), at T = 300 K this field corresponds to $K \gtrsim 2 \times 10^6$ cm^{-1}. Thus, the diffusion mechanism makes an appreciable contribution only at very high values of the space frequency. For photorefractive materials with usual resolution ($K \lesssim 10^4$ cm^{-1}) the diffusion mechanism must not play a substantial role. Neither does it explain the anisotropy of the PR effect [7.23].

7.1.4 The Photoconductivity in an External Field

This mechanism consists in the screening of an external field applied to the crystal by free nonequilibrium carriers. In this case the PR effect is due to a decrease or redistribution of the field in the illuminated region and to the corresponding change in electro-optic effect. The carrier transport mechanism is found to be more effective than in the photovoltaic effect if the electron shift in the external field, $\ell_{\mathscr{E}} = \mu\tau\mathscr{E}_0$, is greater than the asymmetric electron shift $\langle\ell\rangle$ in the AP effect. For instance, in $KTa_{0.65}$

$Nb_{0.35}O_3$ (KTN), at $\mathscr{E}_0$ = 2 kV/cm the shift $\ell_{\mathscr{E}} \simeq 1.5 \times 10^{-3}$ cm, which is five orders of magnitude greater than $\langle \ell \rangle$ in $LiNbO_3$ [6.6,7.23]. In $LiNbO_3$, on the contrary, the values of $\mu\tau$ are so small that the photocurrent for all the fields produced is less than the photovoltaic current. It is possible that the high photorefractive sensitivity of SBN is also associated with the high values of $\mu\tau$.

In this recording mechanism the dark relaxation of the PR effect is associated with the localization of the carriers in the traps and is determined by the Maxwellian relaxation time. The higher $\mu\tau$ and, hence, the higher the photorefractive sensitivity, the faster is the dark relaxation. Photoconductivity and the photovoltaic effect also manifest themselves differently in the mechanism of erasure of the image recorded. When photoconductivity predominates, new illumination of the crystal without an external field leads to its depolarization and returns the crystal to the initial state. If the AP effect predominates, new illumination of the crystal with open electrodes, on the contrary, induces a homogeneous field $\tilde{\mathscr{E}}$ in the crystal volume. In this case, in order to obtain reproducible results erasure must be accomplished with short-circuited electrodes.

7.2 Photorefractive Holographic Recording

Let us consider here, following [7.23], the recording of an elementary hologram on the basis of the PR effect. Suppose the recording medium is a photosensitive electro-optic crystal, which is placed in the plane of intersection of two coherent light beams (the recording and the reference one), as shown in Fig. 7.6. Let the bisector of both beams be perpendicular to the crystal

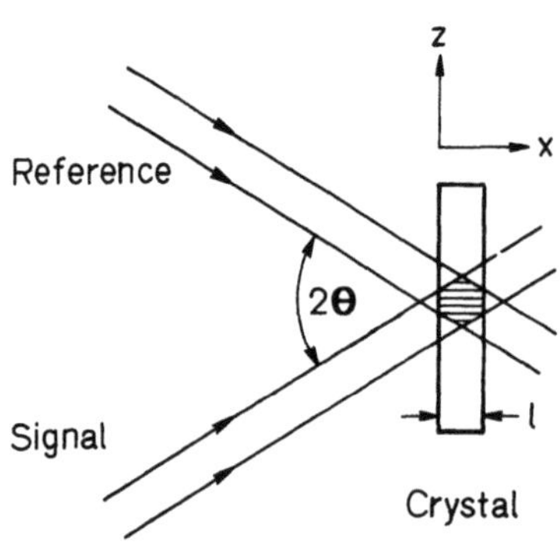

Fig. 7.6. Diagram of photorefractive recording of elementary hologram

surface. Assuming the light intensity I_0 in the two beams to be equal, let us write down the intensity distribution along the z axis as

$$I(z) = 2I_0(1 + \cos|\underline{K}|z) \quad , \tag{7.7}$$

where the spatial frequency of the interference strips is

$$|\underline{K}| = 2\left(\frac{2\pi}{\lambda}\right)\sin\theta \quad . \tag{7.8}$$

Here, λ is the light wavelength in the crystal, and θ is the half-angle between the beams. The light absorption results in the generation of non-equilibrium carriers in the crystal. Assuming that the absorption is weak, i.e. $\alpha\ell \ll 1$ and, hence, the generation of the nonequilibrium carriers is uniform across the thickness of the crystal, ℓ, we write down the distribution of the generation rate along the z axis in the form

$$g(z) = g_0(1 + \cos|\underline{K}|z) \quad , \quad g_0 = \frac{2\gamma\alpha I_0}{\hbar\omega_0} \quad , \tag{7.9}$$

where γ is the quantum yield, and $\hbar\omega_0$ is the photon energy. Let us further consider the two mecahnisms of the PR effect, namely (a) the photoconductivity in the external field; and (b) the influence of nonequilibrium carriers on spontaneous polarization.

a) For the first of these mechanisms, the distribution of the PR effect in the z direction can be obtained from the expression for the current density J, neglecting the diffusion term

$$J = qn\mu\mathscr{E}_0 \quad , \tag{7.10}$$

where the field $\mathscr{E}_0$ in the case of the mechanism at hand (photoconductivity) is the external field applied in the z direction, and n is the concentration of free nonequilibrium carriers. The expression for the current must be supplemented by the Poisson equation (4.3), the continuity equation,

$$\frac{\partial\rho}{\partial t} + \frac{\partial J}{\partial z} = 0 \quad , \tag{7.11}$$

and the kinetics equation

$$\frac{\partial n}{\partial t} = g(z) - \frac{n}{\tau} - \frac{1}{q}\frac{\partial J}{\partial z} \quad . \tag{7.12}$$

Here, τ is the lifetime of the free carriers, which is determined by the recombination mechanism and is independent of n (linear recombination). The set of equations (4.3), (7.10-12) are to be solved for the field distribution in the crystal

$$\mathcal{E}(z) = \mathcal{E}_0 + \Delta\mathcal{E} \quad . \tag{7.13}$$

In [7.23] this solution was obtained under the assumption that the generation rate g satisfying (7.9) is independent of the distribution $\mathcal{E} = \mathcal{E}(z)$ and that consequently $\Delta\mathcal{E}(z) \ll \mathcal{E}_0$ and also under the assumption of a linear relationship $D = \varepsilon\mathcal{E}$.

$$n(z) = \tau g_0\{1 + [1 + (|\underline{K}|\ell_{\mathcal{E}})^2]^{-1/2}\cos(|\underline{K}|z - \varphi) \quad , \tan\varphi = |\underline{K}|\ell_{\mathcal{E}} \ . \tag{7.14}$$

$$\Delta\mathcal{E}(z) = \Delta\mathcal{E}_0 \cos(|\underline{K}|z - \varphi) \quad ,$$
$$\Delta\mathcal{E}_0 = 4\pi\left(\frac{\alpha\gamma W}{\hbar\omega}\right)\frac{q\ell_{\mathcal{E}}}{\varepsilon}\left[1 + (|\underline{K}|\ell_{\mathcal{E}})^2\right]^{-1/2} \quad . \tag{7.15}$$

Here, $W = 2I_0t$ is the density of the total energy incident on the crystal within the exposure time t, $\ell_{\mathcal{E}} = \mu\tau\mathcal{E}_0$ is the carrier shift in the band, and ε is the static dielectric constant of the crystal. The distribution of the PR effect in the z direction is determined by (7.15) and by the electro-optic equation. For instance, for $LiNbO_3$ (point symmetry group 3m), when the optical axis coincides with the z axis, we have, according to (7.4),

$$\Delta n_i = \frac{1}{2}(n_0^3 r_{13} - n_e^3 r_{33})\Delta\mathcal{E}_0 \cos(|\underline{K}|z - \varphi) = \Delta n_{io}\cos(|\underline{K}|z - \varphi) \ , \tag{7.16}$$

where $\Delta n_i = \Delta(n_e - n_0)$. Thus, there is an optical damage distribution in the z direction which reproduces the initial light intensity distribution and represents a recorded elementary phase hologram. The hologram recorded in the form (7.15,16) obeys a reciprocity law analogous to that used in photography [6.11], that is, there is a relationship $\Delta n_i = \Delta n_i(W)$, so that Δn_i

depends only on the product $W = It$. When the reciprocity law is fulfilled, the change in birefringence, Δn_i, and also the diffraction efficiency η (see below) depend exclusively on the total energy $w = \alpha W$ absorbed in the unit volume of the crystal. The fulfillment of the reciprocity law for the photoconductivity mechanism depends on the nature of recombination. The necessary and sufficient condition for the reciprocity law to hold is linear recombination [6.11].

b) If the PR effect is associated with the influence of nonequilibrium carriers on spontaneous polarization P_0, the distribution of Δn_i in the elementary hologram can be obtained from the electro-optic equation recorded in terms of P_0. For instance, for $LiNbO_3$,

$$\Delta n_i = \frac{1}{2}(n_0^3 r_{13} - n_e^3 r_{33}) \frac{\Delta P_0}{\varepsilon} \quad . \tag{7.17}$$

Let us consider the particular case where the change in P_0 is due to the optical recharging of some centers with a concentration M within a time much less than that of Maxwellian relaxation, $t \ll \tau_1$. Then the screening of the depolarization field can be neglected. Equation (7.17) must be supplemented by (2.38)

$$\frac{\Delta P_0}{P_0} = \frac{1}{2}\left(\frac{b}{\beta} - \frac{c}{\gamma}\right) N \quad ,$$

and by the kinetics equations

$$\frac{dn}{dt} = g(z) + k_1 N - \beta_1 n(M - N) - \frac{n}{\tau} \quad , \tag{7.18}$$

$$\frac{dN}{dt} = - k_1 N + \beta_1 n(M - N) \quad . \tag{7.19}$$

Here, N is the concentration of the recharged centers contributing to the change in P_0, and k_1, β_1 are the probabilities of optical expulsion and trapping, respectively. Note that for the relationship between ΔP_0 and N we can use, instead of the thermodynamic relation (2.38), the relations following from the microscopic model of Sect. 3.4. The stationary solution of (7.18,19) is

$$n = \tau g(z) \quad . \tag{7.20}$$

With weak filling of the levels ($\beta_1 n \ll k_1$)

$$N \simeq \frac{\beta_1 M\tau}{k_1} g(z) \tag{7.21}$$

and, as follows from (7.17), (2.38) and (7.21), the distribution of the PR effect in the z direction reproduces the initial light intensity distribution. With strong filling of the levels ($\beta_1 n \gg k_1$) $N \simeq M$, Δn_i does not depend on z, and the PR effect does not reproduce the elementary hologram. The necessary and sufficient conditions for the fulfillment of the reciprocity law in this case at hand are the conditions of quasistationary excitation of the crystal [6.11],

$$n \ll N, \; \frac{dn}{dt} \ll \frac{dN}{dt} \quad . \tag{7.22}$$

The wave signal recorded on the basis of the PR effect can be reconstructed from the elementary hologram by illuminating the crystal with a uniform light beam of wavelength of λ_1 at an angle of θ_1. The diffraction maximum takes place with θ_1 and λ satisfy the Bragg condition

$$\lambda_1 = 2\left(\frac{2\pi}{|\underline{K}|}\sin\theta_1\right) \quad . \tag{7.23}$$

A comparison of (7.23) and (7.8) shows that if the recording and reconstruction of a hologram is accomplished with light of the same wavelength, then $\theta_1 = \theta$. In this case the diffracted light wave reproduces the initial wave exactly although their intensities I_0 and I_1 are not the same, in the general case, and $I_1 \lesssim I_0$. The diffraction efficiency is the ratio $\eta = I_1/I_0$, which, for an elementary phase hologram of the type (7.16), is [7.24]

$$\eta = \ell^{-\alpha\ell} \sin^2\left(\frac{\pi\Delta n_{io}\ell}{\lambda\cos\theta}\right) \quad , \tag{7.24}$$

where the factor $\exp(-\alpha\ell)$ reflects the light absorption by the crystal. Thus, the PR effect can also be observed by measuring the diffraction efficiency η.

We have considered the recording of an elementary hologram corresponding to a single plane wave. But the PR effect can be used for recording an arbitrary hologram corresponding to an arbitrary wave front. This follows from the fact that a complex wave front can, with the aid of a Fourier analysis, be represented as superposition of plane waves, each of them forming an elementary hologram. Thus, the reconstruction of a complex wave front from an arbitrary hologram corresponds to the Fourier synthesis for all the plane waves diffracted from elementary holograms.

7.3 Photorefractive Sensitivity

The photorefractive sensitivity can be defined as $dn_i/d\omega$ [3.21] and analyzed on the basis of the thermodynamic relations (2.14), (2.38), and (7.25), giving

$$n_i = n_e - n_0 = gP_0^2 \quad , \quad \Delta n_i = 2gP_0dP_0 \tag{7.25}$$

where for $g \simeq 1\ \mathrm{cm}^4\ \mathrm{C}^{-2}$ for oxygen-octahedral ferroelectrics [6.8,7.25].

The values of ΔP_0 determined from (7.25) and (2.38) agree if we take for the concentration of nonequilibrium carriers N values typical of other photoferroelectric phenomena. For instance, for the experimental value of Δn_i of 5×10^{-5} for $LiNbO_3$:Fe (Table 7.2), and for $g \simeq 10^8\ \mathrm{cm}^4/\mathrm{C}^2$, (7.25) gives $\Delta P_0 \simeq 4 \times 10^{-9}\ \mathrm{C/cm}^2$. Substituting $\Delta P_0 \simeq 4 \times 10^{-9}\ \mathrm{C/cm}^2$ into (2.38) yields $N \simeq 4 \times 10^{16}\ \mathrm{cm}^{-3}$ for $b \simeq 2.6 \times 10^{-33}\ \mathrm{cm}^2\ \mathrm{V}^{-1}\mathrm{s}^{-1}$ units, $\beta \simeq -7 \times 10^{-13}$ CGSE units, and $c \simeq 0$. The values of b and c were estimated in [1.14] from the expansion of the energy gap E_g in polarization P_0 for $BaTiO_3$ and SBN [see (2.46)]. The experimental value of $\Delta n_i \simeq 1.1 \times 10^{-4}$ for SBN corresponds to a concentration of $N \simeq 2.5 \times 10^{18}\ \mathrm{cm}^{-3}$. The saturation value of $\Delta n_i \simeq 10^{-3}$ in $LiNbO_3$:Fe corresponds to $\Delta P_0 \simeq 2 \times 10^{-7}\ \mathrm{C/cm}^2$ and a concentration of $N \simeq 10^{18}\ \mathrm{cm}^{-3}$. However, these concentrations are much less than those given in [5.17,18;6.6], where for $LiNbO_3$:Cr with a concentration N of Cr^{3+} centers, the change in the dipole moment of the center, $\Delta\mu_0$, and thus $\Delta P_0 = N\Delta\mu_0$, was measured directly. Thus, the saturation value of ΔP_0 was given as $8 \times 10^{-8}\ \mathrm{C/cm}^2$ in [6.6], which, for $\Delta\mu_0 \simeq 10^{-27}$ C.cm, corresponded to a concentration of the excited chromium ions of $N \simeq 8 \times 10^{19}\ \mathrm{cm}^{-3}$. This divergence may be due, firstly, to the different nature of the centers. Secondly it should be borne in mind that, as already noted in Chap. 3, (2,38) describes the change in homogeneous polarization and is, generally speaking, inapplicable to deep localized states.

The PR effect in crystals of $LiNbO_3$:Fe can be observed not only as a change in birefringence, but also as a shift of the intrinsic absorption edgs towards shorter wavelengths. It turns out that the saturation value of $\Delta n_i \simeq 10^{-3}$ corresponds to an increase in the energy gap of $\Delta E_g \simeq 2 \times 10^{-3}$ eV [7.26]. The effect of the shift in the intrinsic optical absorption edge, as well as the change in birefringence, is annealed by heating the crystal to 170°C. An estimate of ΔE_g in the PR effect can be obtained from (2.46):

$$\Delta E_g \simeq aP_0 \Delta P_0 \quad . \tag{7.26}$$

The electron-phonon interaction constant a for $LiNbO_3$ was estimated in [7.26] from electrical absorption. The intrinsic absorption edge of $LiNbO_3$ shifts linearly by $\Delta E_g^{\mathscr{E}} \simeq 2 \times 10^{-3}$ eV > 0 on application of an external field $\mathscr{E}_0 \simeq 10^5$ V·cm^{-1} in the direction of spontaneous polarization P_0. Since $LiNbO_3$ is a pyroelectric with an infinitely large coercive field, the sign of $\Delta E_g^{\mathscr{E}}$ is determined by the sign of the applied field $\mathscr{E}_0$. An estimate yields $a \simeq 0.5 \times 10^{-23}$ CGSE units (for $BaTiO_3$, $a \simeq 0.7 \times 10^{-23}$ CGSE units [1.14]). Substituting this value of a, as well as $P_0 \simeq 70$ μC·cm^{-2} and $\Delta P_0 \simeq 2 \times 10^{-7}$ C.cm^{-2}, into (7.26) yields $\Delta E_g \simeq 0.5 \times 10^{-3}$ eV > 0. This value agrees with the experimental value of ΔE_g corresponding to the saturation of the PR effect in $LiNbO_3$:Fe.

We now discuss the thermodynamic estimate of the photorefractive sensitivity following [6.8].

Let us consider the upper limit of the photorefractive sensitivity, when the entire absorbed light energy w is equal to the change in the free energy of the ferroelectric (2.14) at a constant temperature. Neglecting the elastic stresses in (2.14) and restricting ourselves to the quadratic term away from the Curie point, we have:

$$w = \frac{4\pi}{\varepsilon} PdP \quad , \tag{7.27}$$

where, in the ferroelectric phase, $T < T_1$,

$$P = P_0 + \frac{\varepsilon - 1}{4\pi} \mathscr{E}_0 \quad , \quad \Delta P = \Delta P_0 + \frac{\varepsilon - 1}{4\pi} \Delta\mathscr{E}_0 \quad ; \tag{7.28}$$

and in the paraelectric phase, $T > T_1$

$$P = \frac{\varepsilon-1}{4\pi}\mathscr{E}_0 \quad , \quad \Delta P = \frac{\varepsilon-1}{4\pi}\Delta\mathscr{E}_0 \quad . \tag{7.29}$$

Here, P_0 and $\mathscr{E}_0$ are, respectively, the spontaneous polarization and the external field applied to the crystal. Combining (7.25) and (7.27), we obtain the following expressions for the minimum energy w_{min} and, accordingly, for the upper limit of the photorefractive sensitivity, $(dn/dw)_{max}$:

$$w_{min} = 2\pi\frac{\Delta n_i}{g\varepsilon} \tag{7.30}$$

$$\left(\frac{dn}{dw}\right)_{max} = \frac{1}{2\pi}g\varepsilon \quad . \tag{7.31}$$

Substituting the values of $P_0 \simeq 0.7\ C\cdot m^{-2}$. $\varepsilon \simeq 30$ for $LiNbO_3$ and $\Delta n_i = 10^{-5}$ into (7.30,31), we have $w_{min} \simeq 2\times10^{-2}\ J\cdot cm^{-3}$ and $(dn/dw)_{max} \simeq 0.5\times10^{-3}\ cm^3\ J^{-1}$.

We now consider the actual photorefractive sensitivity in turn for each of the photorefractive mechanisms: the AP effect, the influence of electron excitations on spontaneous polarization, the diffusion mechanism, and the pyroelectric effect. As shown by MICHERON [6.8], in all these cases the photorefractive sensitivity is below $(dn/dw)_{max}$. For the photoconductivity in an external field at sufficiently high values of $\mu\tau$ and of the external field $\mathscr{E}_0$ the photorefractive sensitivity may, on the contrary, greatly exceed $(dn/dw)_{max}$ and reach the sensitivity of silver-halide emulsion layers.

a) The AP Effect. Assuming $J = \varepsilon/4\pi(\partial\mathscr{E}/\partial t)$ in (7.2), the energy $w_{ph} = \alpha I dt$ necessary for inducing a field $\Delta\mathscr{E}$ in the polar direction is given by

$$\Delta\mathscr{E} = 4\pi\frac{k_1}{\varepsilon}w_{ph} \tag{7.32}$$

Using (7.25) and (7.29), we have

$$w_{ph} = \frac{\Delta n_i}{2gP_0k_1} \quad , \tag{7.33}$$

$$\left(\frac{dn}{dw}\right)_{ph} = 2gP_0k_1 \quad , \tag{7.34}$$

Substituting $k_1 \simeq 2.5 \times 10^{-9}$ A·cm (W)$^{-1}$ and $\Delta n_i = 10^{-5}$ into (7.33) and (7.34) for $LiNbO_3$:Fe, we find $w_{ph} \simeq 0.3$ J·cm^{-3} and $(dn/dw)_{ph} \simeq 3 \times 10^{-5}$ cm^3 J^{-1}. Thus, $(dn/dw)_{ph} \ll (dn/dw)_{max}$.

b) The Influence of Electron Excitations on Spontaneous Polarization. Let us assume that the electron excitation of the impurity centers changes their dipole moment by $\Delta\mu_0$. Then the absorption of N photons results in a change of spontaneous polarization ΔP_0.

$$\Delta P_0 = N\Delta\mu \tag{7.35}$$

Denoting by $w_p = N\hbar\omega$ the energy necessary for obtaining the photorefractive effect Δn_i and using (7.25), we have

$$w_p = \frac{h\omega}{2gP_0\Delta\mu}\Delta n_i \quad , \tag{7.36}$$

$$\left(\frac{dn}{dw}\right)_p = \frac{2gP_0\Delta\mu}{\hbar\omega} \quad . \tag{7.37}$$

Let us estimate w_p and $(dn/dw)_p$ for $LiNbO_3:Cr^{3+}$ and $LiTaO_3:Cr^{2+}$, using the data of [6.6]. For $LiNbO_3:Cr^{3+}$, ($\Delta\mu_0 \simeq 0.8$ Debye units, $P_0 \simeq 0.7$ C·m^{-2}), $w_p \simeq 1.0$ J·cm^{-3}, and $(dn/dw)_p \simeq 10^{-5}$ cm^3 J^{-1}. For $LiTaO_3:Cu^{2+}$ ($\Delta\mu_0 \simeq 3$ Debye units, $P_0 \simeq 0.4$ C·m^{-2}), $w_p \simeq 0.25$ J·cm^{-3}, and $(dn/dw)_p \simeq 4 \times 10^{-5}$cm^3 J^{-1}.

c) The Diffusion of Nonequilibrium Carriers. Denoting the diffusion length by L_d and the quantum yield of photoelectrons by γ, we can express the change in polarization ΔP associated with the diffusion of nonequilibrium carriers as

$$\Delta P = \gamma NqL_d \tag{7.38}$$

where N is the number of absorbed photons. Using (7.25) and (7.38), we obtain the energy $w_d = N\hbar\omega$ corresponding to the photorefractive effect Δn_i,

$$w_d = \frac{\hbar\omega}{2gP_0\gamma qL_d}\,\Delta n_i \tag{7.39}$$

$$\left(\frac{dn}{dw}\right)_d = \frac{2g\gamma P_0 qL_d}{\hbar\omega} \quad . \tag{7.40}$$

Substituting into (7.39) and (7.40) the data for $LiNbO_3$:Fe, namely $qL_d \simeq 1.9 \times 10^3$ Debye units [6.40], and assuming that $\gamma = 1$ and $\lambda \simeq 0.5\ \mu m$, we have $w_d \simeq 4.5\ J \cdot cm^{-3}$ and $(dn/dw)_d \simeq 2 \times 10^{-6}\ cm^3\ J^{-1}$. In actual fact, for $LiNbO_3$:Fe the quantum yield is less than unity, as has already been indicated in [6.33]. This is also supported by the measurements in [7.27]. Therefore the photorefractive sensitivity associated with the diffusion mechanism must be still lower.

d) The Pyroelectric Effect. In this case the entire light energy absorbed by the crystal is spent on increasing the crystal temperature, which in turn changes the spontaneous polarization because of the pyroelectric effect.

$$\Delta P_0 = p\Delta T = p\,\frac{w_{py}}{\rho C} \tag{7.41}$$

where p is the pyroelectric coefficient, ρ and C are the density and specific heat capacity, respectively. Using (7.41) and (7.25), we find the expressions for the minimum energy w_{py} and the photorefractive sensitivity $(dn_i/dw)_{py}$,

$$w_{py} = \frac{\rho C}{2pgP_0}\,\Delta n_i \quad , \tag{7.42}$$

$$\left(\frac{dn}{dw}\right)_{py} = \frac{2pgP_0}{\rho C} \quad . \tag{7.43}$$

For $LiNbO_3$, substituting $\rho \simeq 4.64\ g \cdot cm^{-3}$, $C \simeq 0.635\ J \cdot K^{-1}$, and $p \simeq 8.3 \times 10^{-9}\ C \cdot cm^{-2}\ K^{-1}$ [7.28] yields $w_{py} \simeq 0.35\ J \cdot cm^{-3}$ and $(dn/dw)_{py} \simeq 3 \times 10^{-5}\ cm^3\ J^{-1}$.

e) The Photoconductivity in the External Field. Following [6.8], we will show that in an external field it is possible to achieve a photorefractive sensitivity exceeding the value of (7.31). According to this mechanism the energy $w_{pc} = Nu$, which causes a change in birefringence Δn_i, corresponds to transition of N electrons to the band from the impurity centers with energy u (for the intrinsic photorefractive effect $u = E_g$). The shift of the photoelectrons in the band under the action of the external field $\mathcal{E}_0$ leads to a change in polarization

$$\Delta P = Nq\ell_{\mathcal{E}} = Nq\mu\tau\mathcal{E}_0 \quad . \tag{7.44}$$

Using (7.44) and (7.25), we find

$$w_{pc} = \frac{u}{2gPq\mu\tau\mathcal{E}_0}\Delta n_i \quad , \tag{7.45}$$

$$\left(\frac{dn}{dw}\right)_{pc} = \frac{2gPq\mu\tau\mathcal{E}_0}{u} \quad , \tag{7.46}$$

where P satisfies (7.28). Since $P \gg \mathcal{E}_0$, the photorefractive sensitivity is directly proportional to the applied external field. For $LiNbO_3$:Fe, substituting the values of $\mu\tau \simeq 6 \times 10^{-10}$ cm^2 V^{-1}. ($\mu \simeq 15$ cm^2 V^{-1} s^{-1}, $\tau \simeq 4 \times 10^{-11}$ s [6.40]), $\mathcal{E} \simeq 1$ kV/cm, $u \simeq 3.2$ eV, and $\Delta n_i \simeq 10^{-5}$ into (7.45) and (7.46) yields $w_{pc} \simeq 3.8 \times 10^{-3}$ J$\cdot$cm^{-3} and $(dn/dw)_{pc} \simeq 3 \times 10^{-3}$ cm^3 J^{-1}, which is six times the value of (7.31). The authors of [6.33] obtained for $LiNbO_3$:Fe a similar value of $\mu\tau \simeq 10^{-10}$ cm^2 V^{-1}, for which (7.46) and (7.31) coincide. But in [6.33] the mobility in the reduced crystals of $LiNbO_3$:Fe at room temperature was found to be $\mu \simeq 1.1 \times 10^{-3}$ cm^2 V^{-1} s^{-1}. In general, the literature gives contradictory data for μ and τ in $LiNbO_3$, as has already been indicated in Sect. 6.3.

As mentioned above, the PR effect in the external field also takes place in the paraelectric phase and in nonferroelectric photosensitive electro-optic crystals. In this case, substituting (7.29) into (7.45,46) yields

$$w_{pc} = 2\pi \frac{u}{g\mu\tau q\varepsilon\mathcal{E}_0^2}\Delta n_i \tag{7.47}$$

$$\left(\frac{dn}{dw}\right)_{pc} = \frac{1}{2\pi}\frac{g\mu\tau q\varepsilon\mathcal{E}_0^2}{u} \tag{7.48}$$

The photorefractive sensitivity is proportional to the square of the external field. For KTN, substituting $\mu\tau \simeq 1.9 \times 10^{-7}$ cm^2 V^{-1} [7.29], $\varepsilon \simeq 3.6 \times 10^3$, $u \simeq 3$ eV, $\mathscr{E}_0 \simeq 6$ kV/cm, and $\Delta n_i \simeq 10^{-5}$ yields $w_{pc} \simeq 2.5 \times 10^{-3}$ J.cm^{-3} and $(dn/dw)_{pc} \simeq 4 \times 10^{-3}$ cm^3 J^{-1}. Thus, for the indicated value of the external field the photorefractive sensitivity of KTN is close to $LiNbO_3$, although the values of $\mu\tau$ in KTN are much higher. Because of the quadratic dependence of $(dn/dw)_{pc}$ on $\mathscr{E}_0$ in KTN, however, at higher fields the photorefractive effect in KTN is much more sensitive than in $LiNbO_3$.

At present, the optimum values of electro-optic and transport properties have been obtained for nonferroelectric electro-optic photoconductors $Bi_{12}SiO_{20}$ (BSO) and $Bi_{12}GeO_{20}$ (BGO). From the equation for the linear electro-optic effect $\Delta n_i = R\Delta\mathscr{E}_0$, (7.29) and (7.44), we have in this case

$$w_{pc} = \frac{1}{4\pi}\frac{u\varepsilon}{R\mu\tau q\mathscr{E}}\Delta n_i \quad , \tag{7.49}$$

$$\left(\frac{dn}{dw}\right)_{pc} = 4\pi\frac{R\mu\tau q\mathscr{E}}{u\varepsilon} \tag{7.50}$$

In BSO and BGO the photorefractive sensitivity increases linearly with the field. The high values of the electro-optic constant R and the shift $\mu\tau$, and the low values of the dielectric constant ε result in a high photorefractive sensitivity. Substituting for BSO $R \simeq 2 \times 10^{-11}$ cm.V^{-1}, $\varepsilon \simeq 56$, $\mu\tau \simeq 1.2 \times 10^{-7}$ cm^2 V^{-1} [6.8], $u \simeq 2.7$ eV (intrinsic absorption [7.30]), and $\Delta n_i = 10^{-5}$ into (7.49,50), we obtain, for a comparatively low external field $\mathscr{E}_0 \simeq 1$ kV/cm, $w_{pc} \simeq 5.6 \times 10^{-6}$ J.cm^{-3}, $(dn/dw)_{pc} \simeq 1.8$ cm^3 J^{-1}. The energy required for obtaining an appreciable photorefractive effect in BSO (with due allowance for the absorption coefficient) is thus within the range of 100-300 μJ.cm^{-2}, which is close to the photographic sensitivity of silver-halide holographic layers with about the same resolution (~ 100 μJ.cm^{-2}).

7.4 Effect of Photorefraction on Multiphoton Excitation

Until now we have considered photoferroelectric phenomena of an extrinsic or intrinsic nature, when electron excitation occurred on absorption of a single photon. At high light intensity, photoferroelectrics may show multiphoton

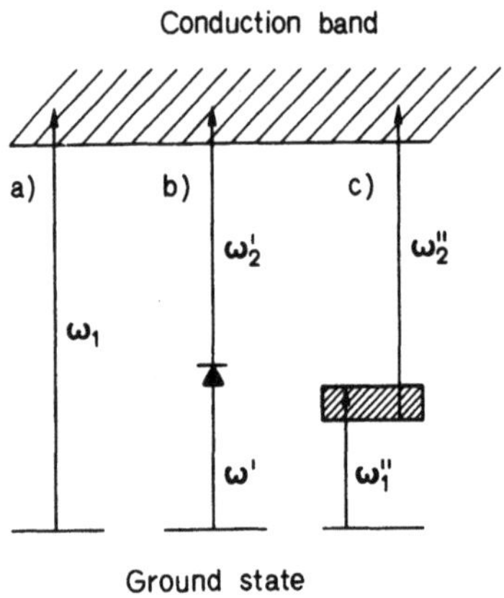

Fig. 7.7. Absorption of light by impurity center [7.23] (a) single-photon absorption; (b) two-photon absorption; (c) two-photon absorption via intermediate state

excitation, when the transition of an electron from the impurity center to the band or from band to band takes place on absorption of several photons. Figure 7.7 shows three possible cases of impurity absorption. Case (a) corresponds to single-photon impurity absorption, when the absorption of a single photon $\hbar\omega$ leads to ionization of the impurity center and transition of the electron to the conduction band. In case (b) the same impurity center is ionized by simultaneous absorption of two photons $\hbar\omega_1'$ and $\hbar\omega_2'$ and the energy $\hbar(\omega_1' + \omega_2')$ corresponds to the optical ionization energy of the center. An impurity center may also be ionized by a third method (c). Here, two-photon absorption results in two-stage ionization of the center. At first, the absorption of the photon $\hbar\omega_1''$ transfers the electrons in the center from the ground to the excited state. Then, on absorption of the photon $\hbar\omega_2''$ the electron shifts from the excited state to the band. The probability of ionization of the center in this latter case naturally depends not only on the photon concentration, but also on the lifetime of the electron in the excited state.

Photorefractive recording by multiphoton excitation has a number of advantages over the case of single-photon absorption discussed throughout the preceeding text. Firstly, the sensitivity of the photorefractive material shifts into the long-wave region of the spectrum. Secondly, investigation into the mechanism of the intrinsic PR effect in broad-band photoferroelectrics is facilitated. Thirdly - and this is particularly important for the application of the PR effect in holography - the reconstruction of holograms in multiphoton photorefractive recording does not result in their degradation. We illustrate this last feature by the diagram in Fig. 7.8. Figure 7.8a reproduces recording of an elementary hologram with the aid of two coherent light

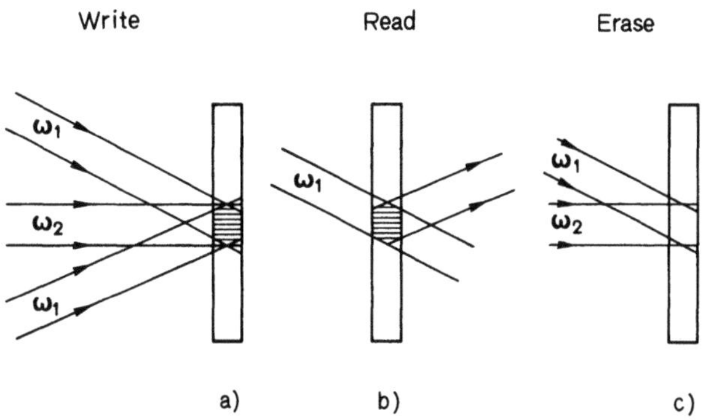

Fig. 7.8. Multiphoton photorefractive recording [7.23] (a) recording of elementary hologram; (b) reconstruction; (c) erasure

beams of frequency ω_1, as shown earlier in Fig. 7.6. The frequency ω_1, however, is selected so that the energy $\hbar\omega_1$ is insufficient for ionizing the impurity centers in a photoferroelectric (the impurity PR effect is considered for definiteness). Therefore, the surface of the photoferroelectric is illuminated simultaneously and uniformly with a light of frequency ω_2 so that the two-photon absorption could lead to the ionization of the centers and hence to the PR effect. Since the illumination of the surface with a light of frequency ω_2 is uniform, the elementary hologram reproduces the interference of the light beams of frequency ω_1, as in Fig. 7.6. It can be seen from Fig. 7.8b that the reconstruction of the recorded hologram is achieved by illuminating the crystal with a coherent light beam of frequency ω_1 at the Bragg angle (7.23). But this single-photon reconstruction does not result in ionization of the centers and in the degradation of the hologram. Therefore, the reconstruction of a hologram recorded in a photoferroelectric on two-photon excitation may be repeated many times without any perceptible damage. The erasure of a recorded hologram is achieved by uniform illumination of the surface with two light beams of frequencies ω_1 and ω_2, i.e. on uniform two-photon excitation (Fig.7.8c).

The PR effect on two-photon absorption in an impurityfree crystal of $LiNbO_3$ was first observed in [7.31]. The crystal was illuminated with a series of ultrashort, highpower, ~ 10 ps pulses of wavelength $\lambda = 0.53$ μm (i.e. in the spectral region where the crystal is transparent). Thus, the excitation corresponded to the case presented in Fig. 6.7b. With a pulse power in the range of 100-700 MW/cm^2 the photon concentration was sufficient for perceptible two-photon intrinsic absorption with a coefficient of $\alpha \simeq 5 \times 10^{-9}$ cm W^{-1}. On two-photon absorption, the light energy w absorbed in unit volume of the crystal is proportional to the square of the incident energy W, i.e.

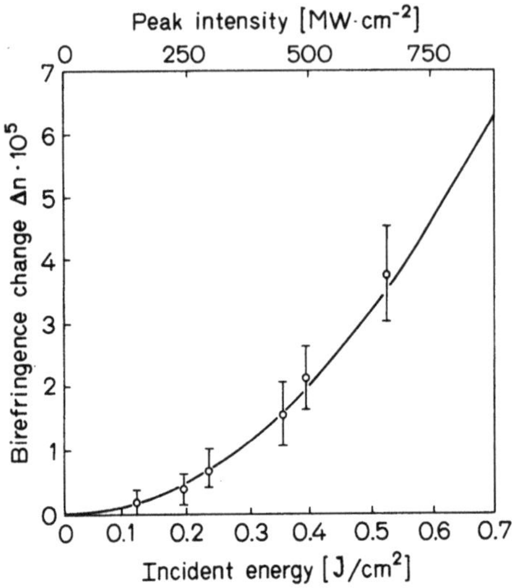

Fig. 7.9. PR effect on two-photon excitation in $LiNbO_3$ as a function of incident energy. Upper scale indicates energy of individual pulse [7.31]

$$w = \alpha I W = \frac{\alpha}{\Delta t} W^2 \quad , \qquad (7.51)$$

where Δt is the exposure time. Since in the PR effect $\Delta n_i \sim w$, the change in birefringence Δn_i is proportional to the square of the incident energy, as can be seen from Fig. 7.9. It appears that the sensitivity of photorefractive recording in $LiNbO_3$ on two-photon absorption is $dn/dw \simeq 2 \times 10^{-5}$ cm^3 J^{-1}, i.e. it is close to the sensitivity of $LiNbO_3$:Fe on single-photon excitation. This is not surprising, since (as shown in Sect.7.3) the sensitivity of photorefractive recording depends on the charge transport mechanism, and not on the excitation conditions.

The PR effect based on two-stage two-photon absorption (Fig.7.7c) in $LiNbO_3:Cr^{3+}$ and $LiTaO_3:Cr^{3+}$ was investigated in [7.23]. Illumination with a wavelength of λ = 650 μm ensured the transfer of the electron in the Cr^{3+} center from the ground state $^4A_2$ to the excited one $^4T_2$. An excited state lifetime of $\tau_e \simeq 10^{-5}$ s was sufficient for the transfer of the excited electron to the conduction band on absorption of the second photon. Two-photon excitation was achieved by means of a ruby laser (λ = 695 μm). For a crystal of $LiNbO_3:Cr^{3+}$ of thickness 0.1 cm with a concentration of Cr^{3+} centers of $\sim 10^{19}$ cm^{-3}, and a series of 50 ns pulses with a total energy of $\sim$ 0.5 $J \cdot cm^{-2}$, a diffraction efficiency of $\eta \simeq 1\%$ was obtained.

For two-photon excitation, as well as for single-photon excitation, in the absence of an external field the photovoltaic effect is evidently the predominant mechanism of photorefractive recording. It is significant that under conditions of two-photon excitation the presence of an intrinsic photovoltaic effect finds the most direct confirmation. But here, too, a reservation is in order. As indicated by recent results [7.32], the intrinsic photorefractive effect in $LiNbO_3$ is associated with the formation, on two-photon absorption, of two new types of impurity color centers, which appear as an impurity absorption band, and change the EPR spectrum. These impurity centers are attributed to the localization of holes near O^{2-} ions and, correspondingly, of electrons near the niobium ions with the formation of Nb^{4+} [7.32]. These centers may play the role, not only of traps during the formation of a space charge, but also of donor centers responsible for the photovoltaic effect, which should, in this case, be considered an impurity effect.

8. Screening Phenomena

As indicated in Sect. 7.1, in some investigations (for instance in [6.16]) the photorefractive effect was assumed to be associated with the screening of the internal field in a ferroelectric (the depolarization field) by non-equilibrium carriers. This assumption is based on the fact that, in principle, an internal macroscopic field always exists even in a short-circuited ferroelectric (Sect.4.1). And although the subsequent investigations showed that the photorefractive effect is mainly due to the photovoltaic effect (at least in $LiNbO_3$:Fe), there are many photoferroelectric phenomena associated with the screening of the internal macroscopic field by nonequilibrium carriers. In [8.1] it was shown that illumination of the ferroelectric SbSI in the intrinsic photosensitivity region in the absence of an external field is accompanied by transitional screening photocurrents and a change in domain structure (the photodomain effect). This group of photoferroelectric phenomena is considered in detail in [1.14]. In this chapter we discuss only some screening phenomena in photoferroelectrics.

8.1 Influence of Nonequilibrium Carriers on Screening of Interfaces

In ferroelectric semiconductors which exhibit a first-order phase transition, such as $BaTiO_3$ and SbSI, the two phases coexist in a certain temperature range near the Curie point, and parts of the ferroelectric phase are single domain [5.11;8.2-4]. The spontaneous polarization is screened near the interface by free or trapped carriers. The problem of the screening of spontaneous polarization at the interface in a ferroelectric semiconductor and of the formation of the periodic structure of interfacts has been considered in Sect. 4.5. In this section we give experimental data on the influence of nonequilibrium carriers on the formation of this structure in SbSI crystals.

Under conditions of strong screening, when the screening energy W_2 is less than the elastic interface energy W_1, plane boundaries are formed between the single-domain parts of the ferroelectric and paraelectric phases [5.11]. For perovskite ferroelectric semiconductors, the density of the elastic interface energy is $W_1 \simeq 30$ erg/cm^2. When the screening energy density is much lower than this value, as in ferroelectric semiconductors $BaTiO_3$ and $KTa_{0.65}Nb_{0.35}O_3$ plane interfacts with (110) orientarion are formed in the region of transition from the cubic to the tetragonal phase. The interface orientation is determined by the anisotropy of elastic stresses and, can be calculated on the basis of the WECHSLER, LIEBERMAN, and READ theory [8.5]. Since, according to (5.4), the screening energy decreases with increasing carrier concentration, it is clear why plane interfaces have been observed only for low-resistance perovskite ferroelectrics ($\rho < 10^3$ Ω cm, $n > 10^{16}$ cm^{-3}) [5.11]. The same fact explains why these interfaces have not been observed in the phase-transition region for high-resistance dielectric crystals of $BaTiO_3$.

Of special interest in connection with the study of the influence of the screening energy on the formation of interfaces is the ferroelectric SbSI. The reason is the presence of nonequilibrium carriers in SbSI and the possibility of varying the concentration of free carriers over a wide range under illumination of a crystal in the intrinsic photosensitivity region. The influence of nonequilibrium carriers on the formation of interfaces in SbSI was investigated in [5.10;8.6,7]. For single crystals of SbSI it was possible, in particular, to study the conditions of formation of interfaces near the Curie point in relation to the carrier concentration in traps, N. At low light intensities, N is low according to (5.7), and thus the screening energy W_2 is high and no plane interfaces are formed. With increasing light intensity the filling of the traps increases, and for some values of N the screening energy is less that the elastic interface energy, $W_2 \lesssim W_1$ in accordance with (5.4). This inequality is the condition for the formation in the crystal of plane interfaces between the single-domain parts of the ferroelectric phase and those of the paraelectric phase. A further increase in light intensity must not change the observed system of interfaces.

Transition to the paraelectric phase in SbSI is accompanied by a discontinuous reduction of the energy gap and a corresponding shift of the intrinsic absorption edge towards longer wavelengths [1.14]. The resulting difference in the absorption of the two phases makes it possible to conduct visual observations of the coexisting phases by illumination with light at a wavelength near the intrinsic absorption edge: the bright regions correspond to the ferroelectric phase, and the dark ones to the paraelectric phase [8.2]. The possibility of optical observation of the coexisting phases in a region far beyond the intrinsic absorption edge is due to a different mechanism. Here, the observation of the structure is associated with the conditions of refraction and reflection at interfaces and crystal faces.

Optical observations of interfaces was carried out in [5.10] with the aid of an infrared microscope equipped with an electron-optical converter. A simicon filter was used for observing crystals in a passing infrared light with a wavelength of $\lambda > 1.1$ μm. The crystal temperature was previously selected so that under additional intrinsic illumination, the crystal showed an interface structure typical of SbSI. A crystal that had been held in the dark for a long time was held at a temperature near the Curie point while it was observed in the microscope under infrared illumination. If no usual interfaces were formed, additional intrinsic illumination was used. At the same time the photocurrent was measured at a constant voltage applied in the [001[direction, and the concentration of free carriers n was estimated for an electron mobility of $\mu \simeq 50\ \mathrm{cm^2\ V^{-1}\ s^{-1}}$. This method was used to measure the value of n corresponding to the light intensity at which plane interfaces with (101) orientation were formed. To estimate the corresponding value of N, measurements of the thermostimulated current were made, with the aid of which and of the well-known technique described in Sect. 5.2 the values of the trap activation energy u, the density of states N_{sm}, and the trap concentration M were calculated. The value of N was estimated from (5.6). Single crystals of SbSI grown from the gas phase, for which the usual interface structure was observed previously by the conventional procedure, were investigated.

The crystals investigated can be divided into two groups according to the results obtained. The first group includes crystals which did not show the usual interface structure during observations in a passing infrared light. For these crystals, interfaces appeared on additional intrinsic illumination. For crystals of the second group, interfaces were observed without additional illumination. Simultaneous measurement of the electrical conductivity showed that htese are low-resistance crystals and, as will be shown below, the condition $W_2 \lesssim W_1$ is satisfied for them due to the high equilibrium carrier concentration.

Figure 8.1 shows microphotographs of a crystal from the first group taken for different illumination intensities and correspondingly different non-equilibrium carrier concentrations. The microphotograph 8.1a, taken in infrared light without additional illumination and corresponding to an equilibrium concentration of $n_0 \simeq 2 \times 10^8\ \mathrm{cm^{-3}}$, shows no interfaces. Plane

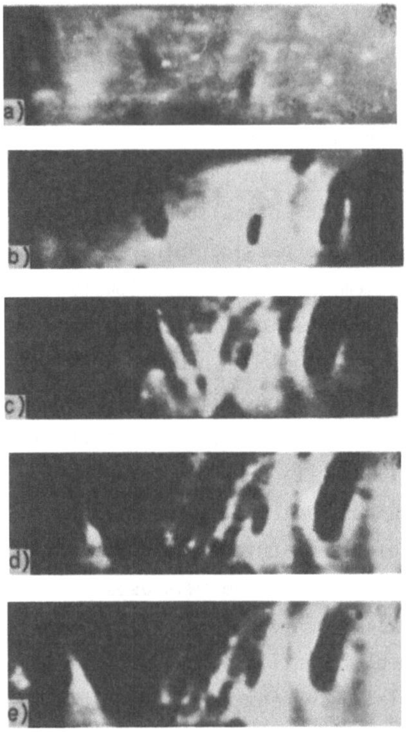

Fig. 8.1. Microphotographs of crystals at various light intensities. a: without additional illumination ($n_0 = 2 \times 10^8\ cm^{-3}$); b,c,d,e: with intrinsic illumination ($n_1 = 4 \times 10^8\ cm^{-3}$; $n_2 = 6 \times 10^8\ cm^{-3}$; $n_3 = 8 \times 10^8\ cm^{-3}$; $n_4 = 10^9\ cm^{-3}$, respectively)

interfaces. Plane interfaces with an orientation typical of SbSI are formed on additional intrinsic illumination (Fig. 8.1c), which corresponds to a nonequilibrium carrier concentration of $n_2 \simeq 6 \times 10^8\ cm^{-3}$. It is seen from Fig. 8.1 that a further increase in illumination intensity does not change the structure observed. Conversely, after prolonged (1 hour) holding of the crystal in the dark the interfaces disappear, and the initial pattern in microphotograph 7.1a is restored.

Measurement by the thermostimulated current method yielded the values $M \simeq 1.4 \times 10^{18}\ cm^{-3}$ and $u \simeq 0.53$ eV (and accordingly $N_{sm} \simeq 8 \times 10^9\ cm^{-3}$) for the same crystal. The concentration $N \simeq 10^{17}\ cm^{-3}$, corresponding to the appearance of interfaces, was determined by substituting these values, as well as $n \simeq 6 \times 10^8\ cm^{-3}$, into (5.6). To estimate the screening energy W_2 by (5.4), it is also necessary to measure the spontaneous polarization P_0 and the dielectric constant ε independently. Here, P_0 and ε denote the parameters corresponding to the separate parts of the ferroelectric phase, ant not the averaged macroscopic parameters of the crystal. Indeed, in the temperature range in which both the phases coexist the crystal is substantially inhomogeneous. In this case the temperature dependences of P_0 and ε, measured for the crystal in the temperature range where it is inhomogeneous, characterize the change in the relative distribution of the phases in the crystal volume, rather than the behavior of one of the phases.

Substituting the values $\varepsilon \simeq 6 \times 10^3$, $P_0 \simeq 14\ \mu C \cdot cm^{-2}$, $\rho = Nq$ into (5.4) for SbSI gives $W_2 \simeq 10\ erg/cm^2$ [5.10]. Using the condition $W_1 \gtrsim W_2$ gives an estimate of the elastic energy of the boundary, W_1, that agrees to an order of magnitude with the value of W_1 for the perovskite ferroelectrics [5.11]. It should be emphasized once again that the uncertainty in the estimate of the parameters P_0 and ε appearing in (5.4) is of fundamental importance, and the above estimate of W_1 is merely an order of magnitude. The estimate of the Debye length, $\ell_D = P_0/\rho$, leads to the value $\ell_D \simeq 6 \times 10^{-4}$ cm. This value agrees fairly well with the Debye length measured directly by the optical method for perovskite ferroelectric semiconductors [5.11] and for SbSI at the boundary of neighboring domains [1.14].

Crystals of the second group, for which interfaces were observed without additional "intrinsic" illumination, had low resistance. The equilibrium carrier concentration, $n \simeq 3.6 \times 10^9\ cm^{-3}$, was measured for one of these crystals. Using the data on the thermostimulated current, the authors determined the values $u \simeq 0.55$ eV, $N_{sm} \simeq 3.5 \times 10^9\ cm^{-3}$, $M \simeq 3 \times 10^{17}\ cm^{-3}$ and calculated the value $N \simeq 1.5 \times 10^{17}\ cm^{-3}$ by (5.6) for the same crystal. It can be seen that without illumination these crystals had a high carrier concentration in the traps, which ensured that the condition $W_2 \lesssim W_1$ held.

As noted in Sect. 4.2, the kinetics of formation of the periodic structure of interfaces is determined by the time required for diffusion-drift equilibrium. This means that the possibility of experimental observation of interfaces in SbSI depends, not only on the screening conditions, but also on the rate of heating or cooling of the crystal during the transition from the ferroelectric to the paraelectric phase, or vice versa. The higher the heating (cooling) rate, the higher must be the nonequilibrium carrier concentration (and hence the shorter the diffusion-drift time) necessary for the appearance of the interface structure. This conclusion was experimentally confirmed for SbSI [8.7]. Figure 8.2 depicts microphotographs of a crystal heated at different rates in the phase transition region. It is seen that the higher the heating rate, the higher is the nonequilibrium conductivity necessary for the appearance of the periodic interface structure.

The experimental data [5.10,8.7] obtained enable one to make some quantitative estimates and compare them with the theoretical values obtained in Sect. 4.2. Thus, according to (4.29) the period of interface structure is equal to the diffusion length $L_D = (D\tau)^{1/2}$, which is assumed to be much greater than the Debye length ℓ_D. If we assume that for SbSI the lifetime of the nonequilibrium electrons and their mobility are $\tau \simeq 10^{-3}$ s and $\mu \simeq 50\ cm^2\ V^{-1}\ s^{-1}$, respectively [1.14], then $L_D \simeq 3 \times 10^{-2}$ cm. If the diffusion

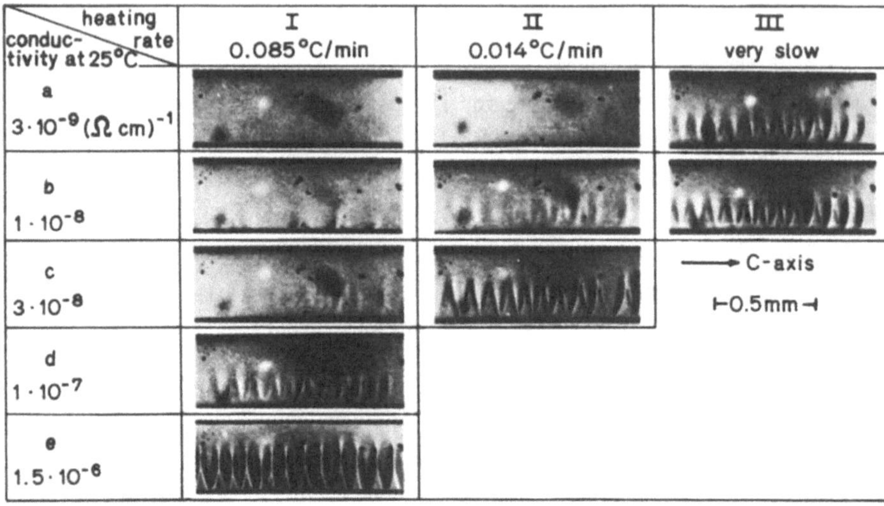

Fig. 8.2. Effect of heating rate on appearance of interfaces [8.7]

length for the holes is close to this value, the theory predicts a value of $\sim 10^{-2}$ cm for the period of the interface structure, which agrees well with the experimental value. Besides, the experimental value of the structure period is found to be much greater than the Debye length in SbSI. Substituting, further, the value $\sigma = W_1 \simeq 20^2$ erg/cm^2, and also the corresponding experimental values of $\Delta E_g \simeq 10^{-2}$ eV [1.14] and L_D into (4.39), we estimate the critical concentration $n_k \simeq 10^{17}$ cm^{-3}, below which no periodic structure appears. We can see that the estimate from (4.39) leads to a critical concentration value which agrees with the above estimates in the order of magnitude.

Such high critical concentration values confirm the assumption that the influence of crystal illumination in the region of its intrinsic photosensitivity on the formation of interfaces is associated with the screening of the single-domain parts of the ferroelectric phase by trapped carriers. The slow dark relaxation of the structure of the coexisting phases and the disappearance of the interfaces after prolonged holding of the crystal in the dark are due to the slow emptying of the traps, when multiple capture evidently plays a substantial role. The slow relaxation of the nonequilibrium conductivity in SbSI also corroborates this mechanism.

8.2 Influence of Screening on Switching Processes

The processes of ferroelectric switching have been widely investigated without consideration of screening [1.13]. However, ferroelectric switching is accompanied by a redistribution of the screening charge, and thus screening charges obviously play a substantial role in switching processes. The surface layers are an integral part of a ferroelectric crystal, being associated in one way or another with the existence of the space-charge screening of the spontaneous polarization. According to various models this charge may either be localized in the surface layer or be distributed at the boundary between this layer and the crystal. The space charge at the surface of $BaTiO_3$ crystals above the Curie point was detected with the aid of pyroelectric measurements by CHYNOWETH [8.8] and with the aid of birefringence measurement by MEYERHOFFER [8.9] and TRIEBWASSER [4.4]. The thickness of the space-charge layer depends on the strength of the field applied to the crystal and the duration of polarization, but the nucleation of antiparallel domains on switching occurs on the surface [1.13]. The formation of the screening space charge and the corresponding increase in the field near the surface layer thus affects the switching mechanism substantially.

The switching mechanism can be characterized by the value of the switching field. In strong fields, the translatory motion of the domain walls predominates. In weak fields, on the contrary, a considerable role in the switching process is played by the lateral motion of the domain walls. The alternation of switching mechanisms in $BaTiO_3$ is also determined by the process of relaxation of the screening space charge. In particular, the space charges reduce the mobility of the 180° domain walls in a high-frequency field without hindering the lateral motion of these walls in low-frequency switching fields. This reduces the high-frequency dielectric constant.

8.2.1 Photoswitching in SbSI

Most of the evidence about the influence of the space charge on the switching process is indirect. In photoferroelectrics such as SbSI the change in the nonequilibrium carrier concentration on illumination makes it possible to observe directly the relationship between ferroelectric switching and screening. The influence of nonequilibrium carriers on switching processes in SbSI has been studied by different methods in various investigations [8.10-12]. We shall now discuss some of these results.

The experimental methods for studying the switching process in SbSI can be divided into two principal groups: (i) direct observation of the motion of the domain walls on switching, and (ii) measurement of the electrical parameters characterizing the polarization value.

In SbSI, only a 180° domain structure is possible, and therefore observation of domains in polarized light is impracticable. The study of Etching figures on the surface of SbSI involves great difficulties. Therefore, in SbSI, switching phenomena and their relationship with screening are naturally best studied by observing the interfaces appearing near the phase transition. As indicated in the preceding section, the charge screening the spontaneous polarization of the single-domain part of the ferroelectric phase is concentrated near the interface. The presence of the screening space charge can be demonstrated by direct optical observation. If a crystal is cooled to the ferroelectric phase, the interfaces naturally disappear, but light bands, evidently due to the shift of the absorption edge under the influence of the space-charge field, remain in their place for a long time.

The switching of parts of the ferroelectric phase must be accompanied by redistribution of the screening charge. This redistribution, in turn, changes the shape of the interface. Thus, observation of the relaxation of interfaces may serve as one of the methods for studying switching processes [8.11].

Optically homogeneous crystals of SbSI with electrodes deposited on their end faces were placed in a thermostated chamber with windows. Two thermostats maintained a constant temperature throughout the chamber with an accuracy to $\pm 0.02^{\circ}$C or set up a preassigned temperature gradient along the [001] axis of the crystal. The light from the monochromator was passed through the parallel (100) faces of the crystal, which could be microphotographed. The periodic structure of interfaces typical of SbSI was observed near the phase-transition temperature, T_1. If the ends of the crystal were maintained at different temperatures, T_2 and T_3, with $T_2 > T_1 > T_3$, then during observation in monochromatic light, corresponding to the absorption edge, one could clearly see ferroelectric and paraelectric regions. When the crystal was observed with light of long wavelength, only the interface was noticeable, as predicted above.

In this section we give the results of investigation into the processes of interface relaxation on pulsed switching of a crystal. The interface was formed due to the temperature gradient along the ferroelectric axis c of the crystal. In constant fields of up to $\mathcal{E} \simeq 800$ V/cm the boundary shape does not change appreciably. With a further increase of the field the boundary smears out, which is evidently due to the smearing of the phase transition in the electric field. The values of the field $\mathcal{E} = V/d$ (V is the voltage applied to the crystal, and d is the interelectrode distance) given here are averaged. Indeed, the screening of spontaneous polarization of the ferroelectric region results in the formation of a layer with a high electric field strength near the interface. For the same reason the measured values of the crystal electrical conductivity σ are averaged. The strength of the field in the ferroelectric region is probably lower, and the electrical conductivity higher than the average values quoted.

The optical method was used for studying the behavior of the interface on reversing the polarity of the voltage applied to the crystal. If the field is low, its reversal does not change the interface. At some critical value $\mathcal{E}_{cr}$ the field reversal results in a smearing out of the boundary with subsequent restoration of the initial pattern within a time τ_1. With an increase in light intensity (and corresponding increase in the photoconductivity of the sample) the critical field increases (Fig.8.3), while the time of formation of the boundary, τ_1, is reduced from several tens of seconds to one second (Fig.8.4). The time τ_1 also depends on the direction of field switchings, and this dependence is not related with the crystal unipolarity.

The interface relaxation mechanism can be visualized as follows. When the critical field is reached, the ferroelectric region of the crystal is switched. The switching time is short, and in SbSI in fields of ~ 200 V/cm it does not exceed milliseconds [1.3]. Illumination does not affect the switching time [8.12]. But the switching process does not end with domain reorientation. It is followed by redistribution of the screening charge in regions of the crystal near the contacts and in the interface region. This results in redistribution of the field in the crystal, which, in turn, may evidently cause additional switching processes. Therefore, on the whole, the switching time is essentially determined by the time of screening of spontaneous

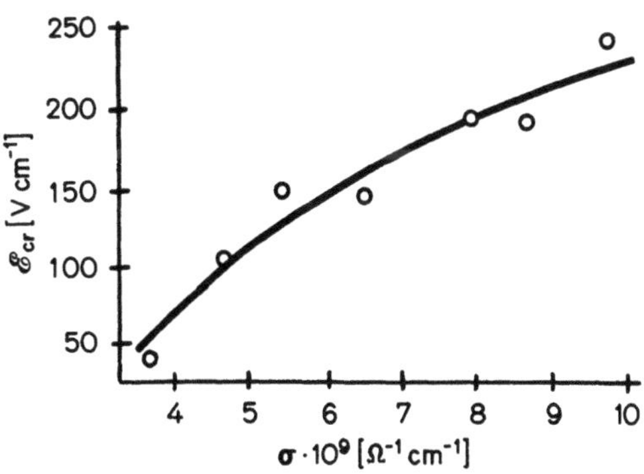

Fig. 8.3. Dependence of critical field causing interface relaxation on photoconductivity

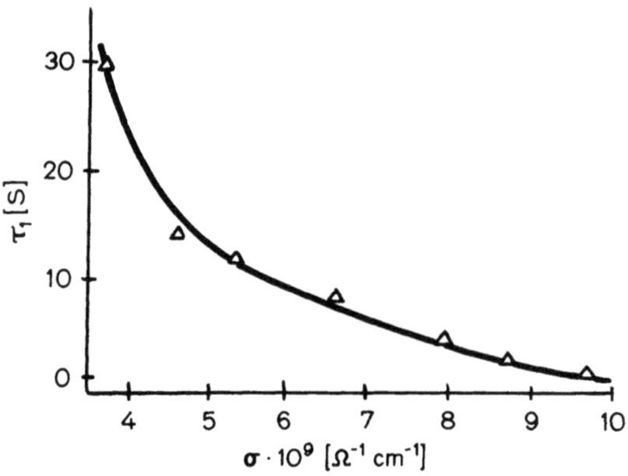

Fig. 8.4. Dependence of interface relaxation time after commutation of electric field on photoconductivity

polarization and the external switching field. Obviously, the interface relaxation time measured in the experiment is equal to the time of formation of the screening space charge at the interface.

It is only after the formation of the screening layer that the interface becomes clear-cut. Indeed, the interface relaxation time coincides in the order of magnitude with the Maxwellian relaxation time in the crystal, $\tau_1 = \varepsilon/4\pi\sigma$. This coincidence could be worse if we took into account that the electrical conductivity near the interface differs from the measured average value. With an increase in photoconductivity, the effective internal field causing the switching decreases due to screening, and this actually leads to an increase in $\mathscr{E}_{cr}$ (Fig.8.3). Since the interface formation time coincides with the Maxwellian time, it is inversely proportional to the photoconductivity (Fig.8.4).

Finally, the dependence of τ_1 on the field switching direction may be associated with the dependence of the electrical conductivity of the screening layer near the interface on the spontaneous polarization direction (the intrinsic field effect). The presence of a layer with a decreased concentration of the principal carriers near the interface results, on the one hand, in an increase of τ_1, and on the other, in a decrease of the effective field strength in the remaining crystal volume. The latter circumstance leads to some shift of the interface and a decrease in the dimensions of the ferroelectric phase previously induced by the field, which is observed experimentally.

Since the switching of parts of the ferroelectric phase is accompanied by redistribution of the screening space charge, the time characteristics of switching depend on the Maxwellian relaxation time. This conclusion was additionally confirmed by investigations of the switching of SbSI in

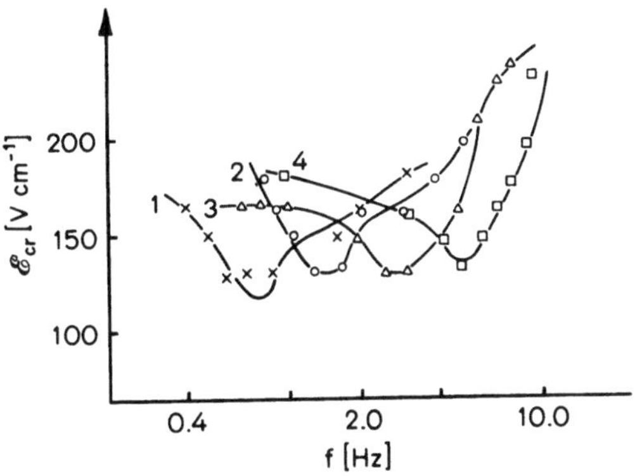

Fig. 8.5. Frequency dependences of critical fields obtained at different levels of crystal illumination for electrical conductivities of 1.1×10^{-8} (1); 2.7×10^{-8} (2); 3.9×10^{-8} (3); 4.5×10^{-8} (4). Ω^{-1} cm^{-1}

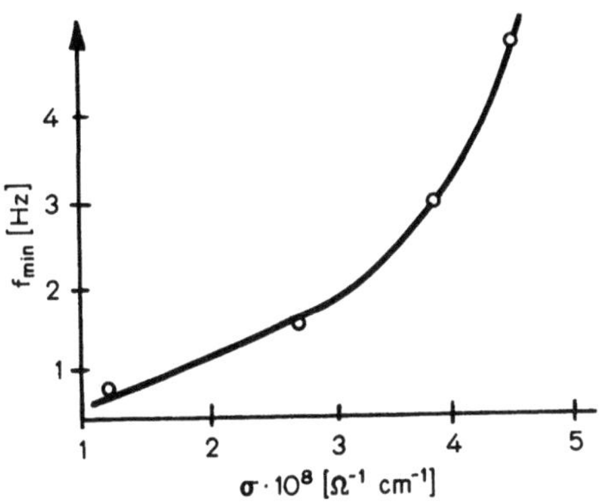

Fig. 8.6. Increase in frequency f_{min} with sample photoconductivity

sinusoidal fields of especially low frequency [8.11]. The SbSI crystals were observed in light; the usual interface structure appeared near the phast transition. An alternating low-frequency field was applied in the direction of the ferroelectric axis of the crystal. Switching of parts of the ferroelectric phase under the action of an alternating field is accompanied by a motion of the interfaces. In these experiments the researchers measured the minimum value of the strength of the applied field $\mathscr{E}_{cr}$ at which interface motion was observed.

The frequency characteristics of the critical fields are presented in Fig. 8.5. It can be seen that the periods corresponding to the minima of curves 1-4 are close to the respective values of the Maxwellian relaxation time. Indeed, for $\varepsilon \simeq 10^4$ and $\sigma \simeq (10^{-9} - 10^{-8})\ \Omega^{-1}$ cm^{-1} the values of $1/\tau_1$ lie in the range of 1-10 Hz. It is well known that deep traps in SbSI are responsible for the slow kinetics of the dark drop of the photocurrent. Accordingly, after the intensive light is switched off the minimum of the frequency characteristic must shift toward the low frequencies, and the kinetics of this shift must correlate with that of the dark drop in the electrical conductivity. This phenomenon was actually observed in [8.11].

The four frequency characteristics in Fig. 8.5 correspond to different intensities of the light used in observation of the crystal, and, accordingly, to different nonequilibrium carrier concentrations. The latter can be determined by measuring the photocurrent corresponding to the same light intensity values. From the data of this figure it follows that with an increase in light intensity, and hence in photoconductivity, the minima of the curves shift toward the higher frequencies. Figure 8.6 shows the dependence of the frequency f_{min}, corresponding to the minima of the curves in the Fig. 7.11, on the crystal photoconductivity.

The results obtained indicate that the frequencies f_{min} can be identified with those corresponding to the Maxwellian relaxation and that the shape of the curves in Fig. 8.5 shows that the process of ferroelectric switching substantially depends on the screening time of the spontaneous polarization.

8.2.2 Photoswitching in PLZT

LAND et al. investigated the influence of photoconductivity on switching processes in the ferroelectric ceramic PLZT [8.13-15]. He showed that illumination of PLZT in the intrinsic photosensitivity region ($E_g \simeq 3.35$ eV) reduces the coercive field. The following technique was used. Transparent indium and tin oxide electrodes were deposited on a ceramic layer of several hundred micrometers. An external field of the order of the coercive one (or slightly smaller) was applied to the electrodes. Then an image ($\lambda \simeq 370$ μm) was projected onto the layer surface through a transparent electrode, and the direction of the external field was reversed simultaneously (switching). The influence of photoconductivity on switching was manifested in the fact that in the illuminated areas the direction of the domains reversed, whereas in the nonilluminated sections the spontaneous polarization retained its former direction. An image corresponding to the two opposite directions of spontaneous polarization appeared in the ceramic layer. This image was observed visually in a reflected of passing light due to the scattering on the surface irregularities associated with surface elastic stresses.

To explain photoswitching in the ceramic PLZT the screening mechanism for SbSI was proposed. When the layer is illuminated, nonequilibrium carriers appear, which are captured by traps and screen the depolarization field. On switching, when the external field reverses its direction, the field strength in the illuminated sections increases, especially near the surface where the screening charge is localized. In [8.15] it was also assumed that a certain role in photoswitching of the creamic PLZT may be played by the field $\tilde{\mathscr{E}}$ generated by the photovoltaic current in the open-circuit regime. As for SbSI crystals, this mechanism was excluded in [8.10-12], because these crystals show low resistance at room temperature, and, according to (6.4), the field $\tilde{\mathscr{E}}$ generated by the photovoltaic current is many orders of magnitude lower than the coercive field. At the liquid-nitrogen temperature, in the highest-resistance crystals of SbSI the field $\tilde{\mathscr{E}}$ may be as high as $\sim$ 100 V/cm and, hence, at low temperatures the photovoltaic effect may influence the photoswitching in SbSI.

8.3 The Photodomain Effect

The influence of illumination and photoconductivity on the equilibrium domain structure, the kinetics of its formation and the properties directly related to the domain structure (such as the pyrocharge and electromechanical hysteresis) constitute what is known as the photodomain effect.

When photoferroelectrics show a low resistance and the field $\tilde{\mathscr{E}}$ generated by the photovoltaic current in the open-circuited crystal is much lower than the coercive field, the contribution of the photovoltaic effect to the photodomain effect can be neglected. This condition holds true, for example, for crystals of SbSI and $BaTiO_3$ near the ferroelectric phase transition. Therefore, the photodomain effect is usually included in the screening phenomena and associated with redistribution of the internal field in the photoferroelectric due to the surface layers, space screening, and recharcing of the surface levels on illumination [1.14].

Here, we discuss some results of experimental investigations into the photodomain effect in barium titanate and antimony sulphoiodide. We first discuss direct observations of the effect, and then a number of properties related to the photodomain effect.

8.3.1 The Photodomain Effect in SbSI and $BaTiO_3$

The photodomain effect was first observed in SbSI crystals [8.1]. Investigation of the photodomain effect in SbSI was carried out by indirect methods, since until recently there were no ways of directly observing the domain structure of this crystal. The influence of illumination on the value of the pyrocurrent in crystals of SbSI and SbSeI was observed in [8.16,17]. Later, the photodomain effect was observed in SbSI when investigating the influence of illumination on Barkhausen pulses [8.10]. Direct observation of the photodomain effect in SbSI was carried out in [8.18], where a single-domain crystal of SbSI was etched before and after illumination. In SbSI, etching reveals 180° domain boundaries parallel to the ferroelectric axis of SbSI. In [8.18] it was shown that illumination of a short-circuited single-domain crystal sharply increases the number of domain boundaries.

The principal results of the experimental investigation into the photodomain effect in crystals of SbSI amount to the following. The spectral distribution of the photodomain effect coincides with that of photoconductivity. The value of the effect increases with the light intensity. The effect is observed only for sufficiently high-resistance, photosensitive crystals. The photodomain effect is observed over a wide temperature range in the ferroelectric region and increases as the Curie point is approached.

The considerable photosensitivity in crystals of $BaTiO_3$ makes possible a detailed investigation into the photodomain effect and particularly into its kinetics, since the symmetry of $BaTiO_3$ crystals permits direct optical observation of the domain structure. The photodomain effect in $BaTiO_3$ was first observed in [5.14]. The effect was investigated for unreduced single crystals of $BaTiO_3$ grown by the Remeika method having a specific resistance of $\sim 10^9 - 10^{10}$ Ω cm. The domain structure was observed in a polarization microscope in the [001] direction through semi-transparent gold electrodes scattered on the (001) face, which made it possible to apply the electric field in the direction of the c axis. The observation was conducted in a monochromatic light, $\lambda = 550$ nm. The crystal was illuminated in the intrinsic region of $350 < \lambda < 450$ nm, where $BaTiO_3$ is photoconductive. In what follows, all measurements made under illumination by long-wavelength, non-photoactive light are termed dark.

The influence of illumination on the kinetics of decay of the c-domain state in a $BaTiO_3$ was investigated in [5.4]. In the course of time a single-domain crystal changes ot a poly-domain one, consisting of a and c domains. Illumination speeds of this process, which is also enhanced by an increase in temperature as the Curie point is approached, as shown in the microphotographs of Fig. 8.7.

As has already been emphasized in Chap. 4, in the case of ideal short-circuiting of a crystal the single-domain state is energetically advantageous. But the existence of surface layers (which is equivalent to the presence of

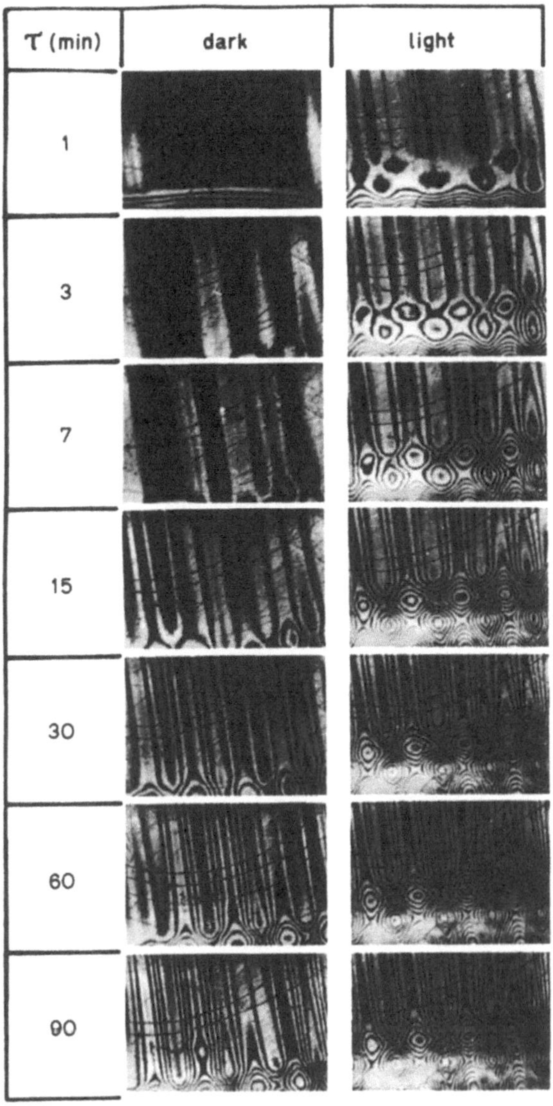

Fig. 8.7. Influence of temperature on photodomain effect in $BaTiO_3$ (a) Formation of new domain in a c-single-domain crystal in the dark and under illumination, T = 110°C.

a dielectric gap) in ferroelectrics, including $BaTiO_3$, results in the appearance of a certain depolarization field in the short-circuited single-domain crystal. The maxima of the depolarization field near the surface causes switching processes. The formation of new domains is rather slow in the dark, since the field near the surface is not high. On illumination the Debye length decreases, and this causes redistribution of the depolarization field, an increase in field strength near the surface, and a decrease

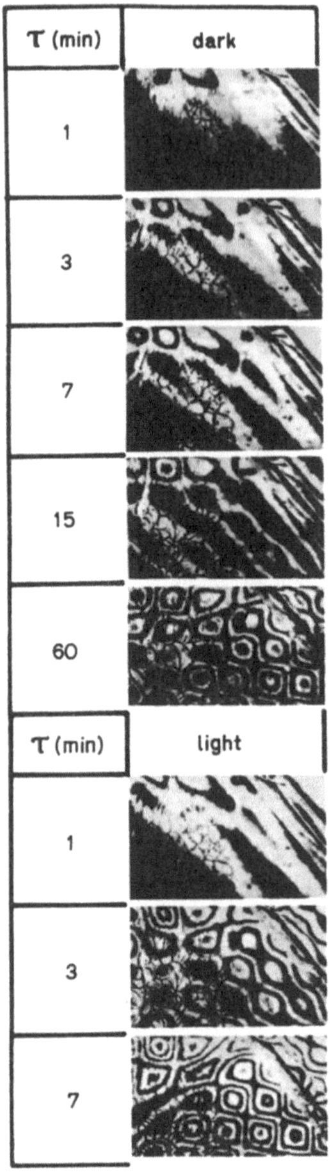

Fig. 8.7. (b) the same processes at T = 125°C (T_1 = 127.5°C)

in the bulk. The increase in field near the surface speeds up the formation of new domains, as illustrated in Fig. 8.7.

The phase transition of a free short-circuited crystal of $BaTiO_3$ results in the formation of the so-called square-lattice Forsbergh structure [8.17]. (It is evident, in particular, in Fig. 8.7.) It was established that on illumination of a crystal with intrinsic light no Forsbergh structure is formed.

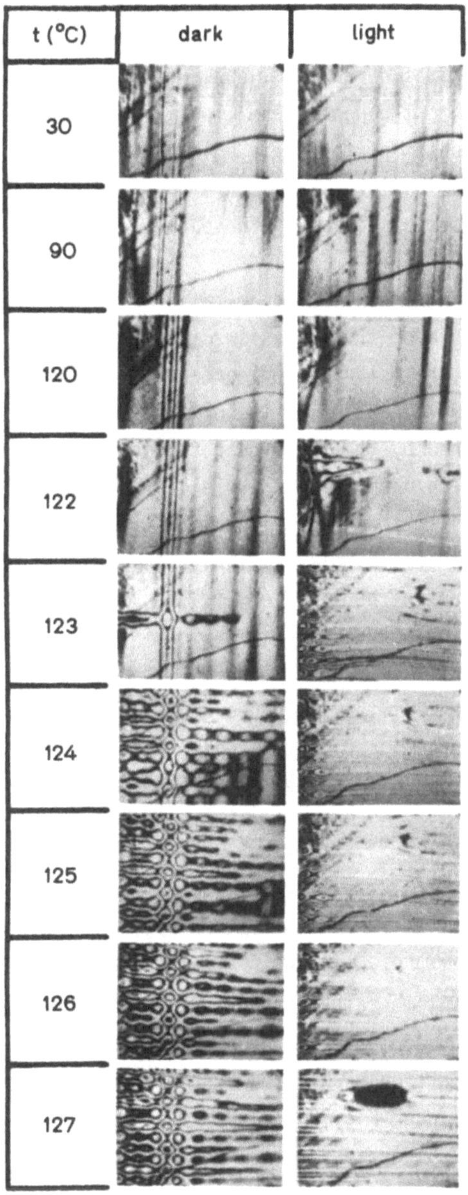

Fig. 8.8. Effect of illumination on equilibrium domain structure of $BaTiO_3$

Figure 8.8 shows successive changes in the equilibrium domain structure of a $BaTiO_3$ crystal with increasing temperature in the dark and under illumination. In the dark, the Forsbergh structure exists in the range 123-127.5°C; the structure does not appear when the crystal is illuminated with intrinsic light.

The mechanism of the influence of illumination on the square-lattice structure is not clear. It is not known, for instance, whether it is associated with screening processes or with the direct influence of nonequilibrium carriers on elastic stresses due to the nonuniform impurity distribution. Figure 8.8 also illustrates the shift of the Curie point toward the lower temperatures under intrinsic illumination. The picture corresponding to 127°C shows that illumination produces a section of the paraelectric phase.

The data on the influence of illumination on the polarization rate of the crystal in the external field also favor the mechanism discussed. According to the existing conceptions of the polarization mechanism in $BaTiO_3$, the process of c-domain formation under the action of the external field in the low-field region is mainly determined by the field near the surface, whereas the process of a-domain formation depends on the field in the crystal volume. The nonequilibrium carriers, which redistribute the field applied to the crystal, must influence the rate of its polarization. Optical observations showed that crystal polarization in the low field (~ 100 V/cm) in the c direction speeds up on illumination. Crystal polarization in the [100] direction, on the contrary, is not accelerated, and in many cases even slows down on illumination. It can be assumed that the influence of illumination on the polarization kinetics is also conditioned by the redistribution of the field in the crystal. Naturally, these effects become more pronounced as the light intensity increases.

The results obtained for $BaTiO_3$ made it possible for the first time to investigate the kinetics of the photodomain effect. This is determined by the longer of the two characteristic times: the Maxwellian relaxation time and the switching time. Calculation of the Maxwellian relaxation time yields a value of 10^{-1} - 1 s for $BaTiO_3$ in the temperature range investigated. Studies into the kinetics of the photodomain effect and polarization under illumination show that these processes are characterized by times of the order ot tens of seconds (see, for instance, Fig. 8.7), i.e. much longer than the Maxwellian time. Thus the switching time determines the kinetics of the photodomain effect.

8.3.2 The Influence of Intrinsic Illumination on Pyrocurrents in SbSI and $BaTiO_3$

In SbSI crystals, in which no optical observation of the domains is possible, the photodomain effect was investigated by measuring the pyroelectric current during the phase transition. It was found that illumination of a single-domain SbSI crystal sharply reduces the value of residual polarization as compared with the spontaneous polarization. A typical case is presented in Fig. 5.2. The photodomain effect recorded by the pyroelectric current should not be confused with the influence on the pyroelectric current of the interfaces formed under intrinsic illumination of the crystal near the Curie point [8.20].

When an SbSI crystal undergoes a phase transition, intrinsic illumination may either decrease the pyroelectric charge of increase it [8.21]. Separate probe measurements showed that the sign of the effect depends on which of the regions near the contact is illuminated. Figure 8.9 shows pyroelectric current curves for one and the same crystal taken during crystal heating. Curve a was obtained for a crystal cooled in the dark. Curve b corresponds to the case where the negative end-face of the crystal was illuminated in the course of the transition from the paraelectric to the ferroelectric phase; curve c was obtained under the same conditions, except that the positive end-face was illuminated. These results show that illumination of the negative end-face decreases the pyroelectric charge and, hence, the unipolarity of the crystal and illumination of the positive end-face increase them. Investigation into the spectral distribution of these effects showed the presence of two maxima: an intrinsic one at $\lambda \simeq 620$ nm ($E_g \simeq 2$ eV for SbSI) and an extrinsic one at $\lambda \simeq 720$ nm.

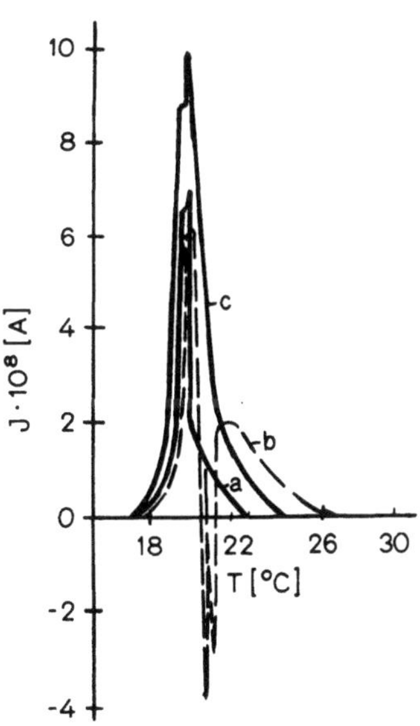

Fig. 8.9. Temperature dependence of pyrocurrent in natural polarization regime [8.22]

These results do not contradict the decrease in the pyroelectric charge of a single-domain crystal on its illumination in the ferroelectric region, and they can be explained on the basis of the mechanism of optical recharging of the surface levels in the paraelectric region. Assume that the natural unipolarity of the crystal is due to the unequal charge of the surface levels on the opposite end-faces of the crystal in the paraelectric region. For definiteness, let the positive charge be localized on both end-faces, the charge on end-face 1 exceeding the charge on end-face 2 (Fig.8.10). Since the field at surfacd 1 is larger than at surface 2, and nucleation and domain growth is determined by the field near the surface and begins from the surface [1.13], during the transition of the crystal from the paraelectric to the ferroelectric phase, ferroelectric polarization P_0 arises in it; its direction is indicated by an arrow in Fig. 8.10 (natural unipolarity or natural polarization). Naturally, in the ferroelectric phase the band curvature and the field distribution in the crystal no longer correspond to Fig. 8.10, because they are also affected by the screening of the resulting natural polarization.

To explain the influence of light on unipolarity, let us assume that under illumination of the crystal in the paraelectric region (or during the transition through the Curie point) the positively charged surface levels of both end-faces capture nonequilibrium electrons, and their charge decreases. (Accordingly, nonequilibrium holes are captured by levels in the volume and decrease the space charge.) Then illumination of surface 2 increases the difference between the charges on the opposite end-faces and thus increases the unipolarity and natural polarization. When surface 1 is illuminated, on the contrary, the difference between the charges and, hence, the pyroelectric charge being measured decreases. On illumination of a naturally polarized crystal in the ferroelectric region the situation is more complicated, because it also depends on the influence of nonequilibrium carriers on polarization

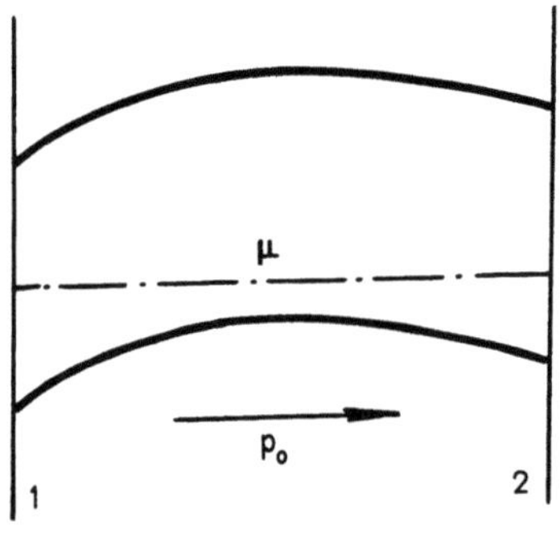

Fig. 8.10. Band curvature at crystal end-faces in paraelectric

screening. Under these conditions the photodomain effect may be accompanied either by an increase or by a decrease in unipolarity [8.21]. If the impurity maximum at $\lambda = 0.72\ \mu m$ corresponds to transition of electrons from the valence band to the surface layers in SbSI, the presence of two maxima in the spectral distribution of the photodomain effect can also be understood. Thus, the probe investigation of the photodomain effect in the natural polarization regime enables one to draw conclusions as to the field distribution in the paraelectric region. The nature of the field distribution in a short-circuited ferroelectric can also be investigated with the aid of the short-circuit photocurrent measured in the probe regime; we will discuss this in the next section.

8.4 Screening and Short-Circuit Photocurrents

The photodomain effect indicates the presence of an internal electric field in a short-circuited ferroelectric of both one or more domains. The existence of an internal field in a short-circuited photoferroelectric can be checked additionally by observing the short-circuit photocurrent J_{sc} associated with the nonequilibrium carrier drift in the internal field of the crystal. Moreover, in probing a crystal with light, this method can be used to investigate the distribution of the internal field along the spontaneous polarization axis, which can be compared with the theoretical one obtained for a single-domain crystal with a free surface (see Chap.4).

The short-circuit photocurrent J_{sc}, which is due to screening, should not be confused with the photovoltaic current J, which was discussed in Chap. 6. The photovoltaic current J is stationary and is observed in the absence of an internal macroscopic field $\tilde{\mathscr{E}}$ in a ferroelectric. The short-circuit photocurrent is a transient current and is due to the screening of the internal field $\tilde{\mathscr{E}}$ by nonequilibrium carriers. When the light is switched on the short-circuit photocurrent reaches a maximum and then falls off to zero or to a stationary value corresponding to the photovoltaic current. The value of J_{sc} is usually estimated by the maximum. Hence it is clear that separation of the short-circuit photocurrent J_{sc} from the pyrocurrent, associated with the heating of the crystal under illumination, can be achieved easily enough only for ferroelectrics with a high photosensitivity, such as SbSI.

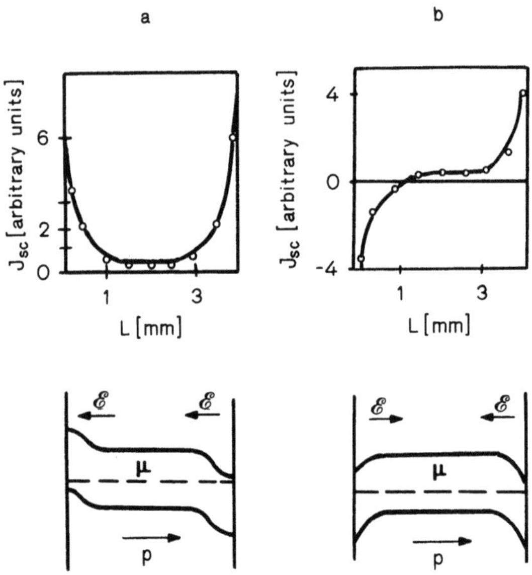

Fig. 8.11. Two types of distribution of J_{SC} for naturally polarized SbSI crystal during probing with light beam along c axis [8.22]. (a) distribution of J_{SC} and band curvature for a-group crystals; (b) distribution of J_{SC} and band curvature for b-group crystals

The first observation of a short-circuit photocurrent in single-domain crystals of SbSI was made in [8.1]. The distribution of the internal field in SbSI crystals was obtained in [8.22] from measurements of the photocurrent J_{sc} in the light probing regime. The crystal was cooled and transferred to the ferroelectric phase (without preliminary polarization), where the photocurrent J_{sc} was probed with light and recorded during the shifting of the light probe between the electrodes in the direction of the c axis. The results of these measurements showed that SbSI has two types of internal field distribution, a and b, which are presented in Fig. 8.11. The lower the unipolarity of the crystal, the more often it can be referred to group b and, the higher the unipolarity, the more often the crystal belongs to group a. After polarization in an external field exceeding the coercive one, a b-type crystal, as a rule, has the a-type distribution of the internal field. If this crystal is heated to a temperature above the Curie point and cooled, it will again reveal the b-type distribution in the ferroelectric phase.

All these and other results indicate that the nature of the internal field distribution in SbSI is determined by two mechanisms. The first mechanism is associated with the screening of ferroelectric polarization by the charge localized at the levels in the volume and on the surface. This mechanism causes bending of the bands and results in an internal field distribution shown in Fig. 8.11a (type a). The a-type distribution is undoubtedly due to the participation of the surface levels in screening, since degeneracy is

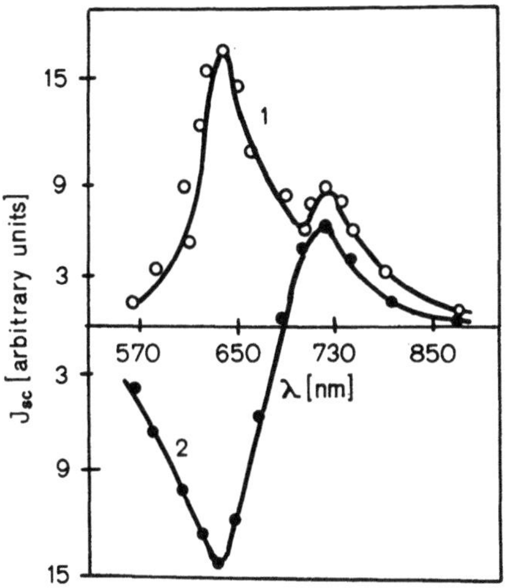

Fig. 8.12. Spectral dependence of J_{SC} for a-group (1) and b-group (2) crystals

not observed in SbSI. The second mechanism is associated with the charge of the surface levels; it causes bending of the bands and determines the distribution of the field near the surface in the paraelectric region. In the absence of ferroelectric polarization this mechanism always leads to the b-type distribution. As mentioned in Sect. 8.3, the same mechanism is responsible for the crystal unipolarity.

The above conclusions are confirmed by the spectral distribution of the photocurrent J_{sc} taken for the opposite end-faces of the crystal (see Fig. 8.12). Curve 1 represents the spectral distribution of J_{sc} for a-type crystals, and curve 2, for b-type crystals. The a-type spectral distribution is the same for both end-faces (curve 1) and reveals both an intrinsic maximum at $\lambda \simeq 620$ nm and an impurity maximum at $\lambda \simeq 720$ nm, which the authors of [8.22] attribute to transitions of the electrons from the valence band to the surface levels. In b-group crystals, the spectral distributions of J_{sc} are different for the opposite end-faces. On the positive end-face the spectral distribution is the same as in an a-type crystal (curve 1). On the negative end-face there is a spectral distribution with a change in the sign of J_{sc} on transition from the long-wavelength maximum to the short-wavelength one. This is attributed to the decrease in the charge of the surface levels under illumination by light of wavelength $\lambda \simeq 720$ nm and hence to the change in the sign of the curvature of the bands near the surface.

References

1.1 B.M. Vul: Dokl. Akad. Nauk SSSR. 43, 308 (1944)
1.2 E. Fatuzzo, G. Harbeke, W.J. Merz, R. Nitsche, H. Roetschi, W. Ruppel: Phys. Rev. 127, 2036 (1962)
1.3 R. Nitsche, H. Roetschi, P. Wild: Appl. Phys. Lett. 4, 210 (1964)
1.4 M.E. Lines, A.M. Glass: *Principles and Applications of Ferroelectrics and Related Materials* (Clarendon Press, Oxford 1977)
1.5* G.A. Smolensky, V.A. Bokov, V.A. Isupov, N.N. Krainik, R.E. Pasynkov, M.S. Shur: *Ferroelectrics and Antiferroelectrics* (Nauka, Moscow 1971)
1.6* V.G. Vaks: *Introduction to Microscopic Theory of Ferroelectrics* (Nauka, Moscow 1973)
1.7* E.V. Bursian: *The Nonlinear Crystal, Barium Titanate* (Nauka, Moscow 1974)
1.8 E. Fatuzzo, W.J. Merz: *Ferroelectricity*, (North-Holland, Amsterdam 1967)
1.9 J.C. Burfoot: *Ferroelectrics: An Introduction to the Physical Principles* (Princeton, New Jersey 1967)
1.10 R. Blinc, B. Žeks: *Soft Modes in Ferroelectrics and Antiferroelectrics* (North-Holland, Amsterdam 1974)
1.11 E.J. Samuelsen (ed.): *Structural Phase Transition and Soft Modes* (Universitetvorleiget, Oslo 1971)
1.12 S.B. Lang: *Sourcebook of Pyroelectricity* (Gordon and Breach, New York 1974)
1.13 F. Jona, G. Shirane: *Ferroelectric Crystals* (Pergamon, Oxford 1962)
1.14* V.M. Fridkin: *Ferroelectric Semiconductors* (Nauka, Moscow 1976)
2.1 V.M. Fridkin, L.M. Belyaev, A.A. Grekov, A.I. Rodin: J. Phys. Soc. Jpn. 28, Suppl. 448 (1970)
2.2 F. Micheron, J.C. Trotier: Ferroelectrics 8, 441 (1974)
2.3 G. Chanussot: In *Ferroelectrics: 4th International Meeting on Ferroelectricity* (Leningrad, 1977). In press
2.4 G.A. Smolensky, V.M. Fridkin, A.A. Grekov: *Proc 1972 Rostov-on-Don Symp. Semiconducting Ferroelectrics*, Ferroelectrics 6, 1/2 (1973)
2.5 V.M. Fridkin: JETP Lett. 3, 252 (1966)
2.6 R.E. Pasynkov: Izv. Akad. Nauk SSSR, Ser. Fiz. 34, 2466 (1970)
2.7 R.E. Pasynkov: Ferroelectrics 6, 19 (1973)
2.8 L.D. Landau, E.M. Lifshitz: *Statistical Physics*
2.9 V.L. Ginsburg: Usp. Fiz. Nauk 38, 490 (1949)
2.10 V.L. Ginsburg: Usp. Fiz. Nauk 77, 621 (1962)
2.11 J. Grindlay: Introduction to the Phenomenological Theory of Ferroelectricity (Pergamon, Oxford 1970)
2.12 W.J. Merz: Phys. Rev. 95, 690
2.13 R. Kern: J. Phys. Chem. Solids 23, 249 (1962)
2.14 G. Harbeke: J. Phys. Chem. Solids 24, 957 (1963)
2.15 H.M. James: Photoconductivity Conference (Atlantic City, N.Y. 1954)

*References to works in Russian whose titles are given here in translation are marked with an asterisk

2.16 H. Brooks: Adv. Electron. Electron Phys. 7, 117 (1955)
2.17 R.W. Keyes: Solid State Phys. 11, 179 (1960)
2.18 J.C. Slater: *Introduction to Chemical Physics* (McGraw-Hill, New York 1939) p. 216
2.19 N.P. Mott, R.W. Gurney: *Electronic Processes in Ionic Crystals* (Clarendon Press, Oxford 1940)
2.20 I.B. Levinson, E.I. Rashba: Usp. Fiz. Nauk 108, 385 (1972)
3.1 W. Cochran: Adv. Phys. 9, 387 (1960)
3.2 J.J. Hallers, W.T. Caspers: Phys. Stat. Sol. 36, 587 (1969)
3.3 K. Kreher: Phys. Lett. A30, 384 (1969)
3.4 K. Kreher: Wiss. Z., Karl-Marx-Univ. Leipzig, Math.-Naturwiss. Reihe 20, 287 (1971)
3.5 T. Nattermann: Phys. Stat. Sol (b) 51, 395 (1972)
3.6* S.M. Ryvkin: *Photoelectric Phenomena in Semiconductors* (Fizmatgiz, Moscow 1963)
3.7 J.M. Ziman: *Principles of the Theory of Solids* (Cambridge University Press, Cambridge 1964)
3.8 E. Pytte: Phys. Rev. B5, 3758 (1972)
3.9 R. Migoni, H. Bilz, B. Reiser, D. Bäuerle: In *Ferroelectrics: 4th International Meeting on Ferroelectricity* (Leningrad, 1977) In press
3.10 K. Sinha, A. Sinha: Indian J. Pure Appl. Phys. 2, 91 (1964)
3.11 I.B. Bersuker: Phys. Lett. 20, 589 (1966)
3.12 N. Kristoffel, P. Konsin: Ferroelectrics 6, 3 (1973)
3.13 G. Chanussot: Ferroelectrics 8, 671 (1974)
3.14 A.P. Levanyuk, V.V. Osipov: Izv. Akad. Nauk SSSR, Ser. Fiz. 39, 686 (1975)
3.15 I.M. Lifshitz: Usp. Fiz. Nauk 83, 617 (1964)
3.16 M.A. Krivoglaz: Fiz. Tverd. Tela, Leningrad 11, 2230 (1969)
3.17 M.A. Krivoglaz: Fiz. Tverd. Tela, Leningrad 12, 1705 (1970)
3.18 E.A. Galashin: Dokl. Akad. Nauk SSSR 171, 366 (1966)
3.19 I. Tyndall: Proc. R. Soc. London 17, 92 (1869)
3.20 C.T.R. Wilson: Phys. Trans. R. Soc. London, Ser A: 192, 403 (1899)
3.21 A.M. Glass, D. von der Linde, D.H. Auston, T.H. Negram: J. Electron Mater. 4, 915 (1975)
3.22 M.T. Mulcahy, E. Kuffel: Proc. Phys. Soc. 80, 1333 (1962)
3.23 I. Förster, K. Kasper: Zs. Electrochem. 59, 975 (1955)
3.24 E.A. Galashin, S.S. Dorofeyev: In *Mekhanism i kinetika kristallizatsii* (Crystallization Mechanism and Kinetics), (Nauka i tekhnika, Minsk 1969) p. 483
3.25 H.M. Papec, A.C. Montefinale, T.V. Zanidzki: Nature 203, 1343 (1964)
3.26 A.V. Ginzberg, A.D. Sablin-Yavorsky, V.M. Fridkin: JETP Lett. 12, 22 (1970)
3.27 C.B. Van den Berg, I.E. Van Delden, I. Bouman: Phys. Stat. Sol. 36, K89 (1969)
3.28 P. Wachter, P. Weber: Solid State Commun. 8, 1133 (1970)
3.29 V.A. Izvozchikov, V.A. Bordovsky: In *Poluprovodnikisegnetoelektriki* (Semiconductor-Ferroelectrics) (RGU, Rostov-on-Don 1973) p. 152
4.1 M. DiDomenico, S.H. Wemple: Phys. Rev. 155, 539 (1967)
4.2 E.V. Chensky, V.B. Sandomirsky: Fiz. Tekh. Poluprovodn. [Sov. Phys. - Semicond.] 3, 857 (1969)
4.3 E.V. Chensky: Fiz. Tverd. Tela (Lenningrad) 12, 586 (1970)
4.4 S. Triebwasser: Phys. Rev. 118, 100 (1960)
4.5 A.I. Larkin, D.E. Khmelnitsky: Zh. Eksp. Teor. Fiz. 55, 2345 (1968)
5.1 W. Känzig: *Ferroelectrics and Antiferroelectrics* (Academic Press, New York 1957)
5.2 L.M. Belyaev, I.I. Groshik, V.A. Lyakhovitskaya, V.N. Noscov, V.M. Fridkin: JETP Lett. 6, 481 (1967)
5.3 B.P. Grigas, I.P. Grigas, R.P. Belyatskas: Fiz. Tverd. Tela (Leningrad) 9, 1532 (1967)

5.4 S. Ueda, I. Tatsuzaki, J. Shindo: Phys. Rev. Lett. 18, 453 (1967)
5.5 M.D. Volnyansky, A.Yu. Kudzin, A.N. Sukhinsky: Fiz. Tved. Tela (Leningrad) 15, 2819 (1973)
5.6 G. Godefroy, P. Jullien, L. Cai: Ferroelectrics 13, 309 (1976)
5.7 R.H. Bube: *Photoconductivity of Solids* (Wiley, New York 1960)
5.8 R.H. Bube, S. Thomsen: J. Chem. Phys. 23, 15 (1955)
5.9 V.A. Alekseyeva, E.G. Lansberg: Fiz. Tverd, Tela (Leningrad) 8, 3138 (1966)
5.10 I.I. Groshik, V.M. Fridkin: Fiz. Tverd. Tela (Leningrad) 10, 2878 (1968)
5.11 M. DiDomenico, S.H. Wemple: Phys. Rev. 155, 539 (1967)
5.12 A.A. Grekov, A.I. Rodin, V.M. Fridkin: Fiz. Tverd. Tela (Leningrad) 12, 3643 (1970)
5.13 G.L.F. Garlick, A.F. Gibson: Proc. R. Soc. London 60, 574 (1948)
5.14 T.R. Volk, A.A. Grekov, N.A. Kosonogov, A.I. Rodin, V.M. Fridkin: Kristallografiya 16, 241 (1971)
5.15 T.R. Volk, A.A. Grekov, N.A. Kosonogov, V.M. Fridkin: Fiz. Tverd. Tela (Leningrad) 14, 3216 (1972)
5.16 L. Benguigui: Solid State Commun. 7, 1245 (1969)
5.17 A.M. Glass, D.H. Auston: Opt. Comm. 5, 45 (1972)
5.18 A.M. Glass, D.H. Auston: Ferroelectrics 7, 187 (1974)
5.19 D.H. Auston, A.M. Glass: Appl. Phys. Lett. 20, 398 (1972)
5.20 I. Tatsuzaki, K. Itoh, S. Ueda, I. Shindo: Phys. Rev. Lett, 17, 198 (1966)
5.21 A.A. Grekov, E.D. Rogach, L.N. Syrkin: Fiz. Tverd. Tela (Leningrad) 14, 2768 (1972)
5.22 V.M. Fridkin, K. Gulyamov, V.A. Lyakhovitskaya, V.N. Nosov, N.A. Tikhomirova: Fiz. Tverd. Tela (Leningrad) 8, 1907 (1966)
5.23 A.M. Zavyalova, P.L. Zaks, L.N. Syrkin: Fiz. Tverd. Tela (Leningrad) 12, 1580 (1970)
5.24 N.A. Abramov, V.V. Voronov, Yu.S. Kuzminov: In *Ferroelectrics: 4th International Meeting on Ferroelectricity* (Leningrad, 1977) In press
5.25 V.L. Ginsburg: Dokl. Akad. Nauk SSSR 105, 240 (1955)
5.26 G. Chanussot, V.M. Fridkin, G. Godefroy, B. Jannot: Appl. Phys. Lett. 31, 3 (1977)
5.27 G. Chanussot, V.M. Fridkin, G. Godefroy, B. Jannot: In *Ferroelectrics: 4th International Meeting on Ferroelectricity* (Leningrad, 1977) In press
5.28 G. Chanussot, V.M. Fridkin: Appl. Phys. Lett. 1978, in press
6.1* E.I. Adirovich: Photoelectronic Phenomena in Semiconductors and Optical Electronics (Fan, Tashkent 1972)
6.2 A.M. Glass, D. von der Linde, T.J. Negran: Appl. Phys. Lett. 25, 233 (1974)
6.3 A.A. Grekov, M.A. Malitskaya, V.D. Spitzina, V.M. Fridkin: Kristallografiya 15, 500 (1970)
6.4 T.R. Volk, A.A. Grekov, N.A. Kosonogov, V.M. Fridkin: Fiz. Tverd. Tela (Leningrad) 14, 3216 (1972)
6.5 V.M. Fridkin, B.N. Popov, K.A. Verchovskaya: Phys. Stat. Sol. (a), 39, 139 (1977)
6.6 A.M. Glass, D. von der Linde: Ferroelectrics 10, 163 (1976)
6.7 P. Günter, P. Micheron: Ferroelectrics 18, 27 (1978)
6.8 F. Micheron: Ferroelectrics 18, 153 (1978)
6.9 A.P. Levanyuk, V.V. Osipov: Izv. Akad. Nauk SSSR, Ser. Fiz. 41, 752 (1977)
6.10 Y. Yasojima, Y. Ohmozi, Y. Inuishi: Tech. Report. Osaka University 22, 575 (1973)
6.11 V.M. Fridkin: *The Physics of the Electrophotographic Process* (Focal Press, London 1973)
6.12 W.T.H. Koch: Thesis, University of Karlsruhe (1975)
6.13 W.T.H. Koch, R. Munser, W. Ruppel, P. Würfel: Sol. State Comm. 17, 847 (1975)

6.14 W.T.H. Koch, R. Munser, W. Ruppel, P. Würfel: Ferroelectrics 13, 305 (1976)
6.15 P. Günter, U. Flückiger: Ferroelectrics 13, 297 (1976)
6.16 F.S. Chen: J. Appl. Phys. 40, 3389 (1969)
6.17 I.P. Kaminov, E.H. Turner: Proc. IEEE 54, 1374 (1966)
6.18 P. Günter: Opt. Commun. 11, 285 (1974)
6.19 E. Wiesendanger: Ferroelectrics 6, 263 (1974)
6.20 L.A. Ivanova, V.I. Popolitov, S.Yu. Stefanovich, A.N. Lobachev, Yu.N Venevtsev: Kristallografiya 19, 573 (1974)
6.21 P.V. Ionov: Fiz. Tverd Tela (Leningrad) 15, 2827 (1973)
6.22 H. Kurz: Ferroelectrics 13, 291 (1976)
6.23 V.M. Fridkin, B.N. Popov: Ferroelectrics 18, 165 (1978)
6.24 K. Ohi, I. Takeda, Yu. Ohara: Izv. Akad. Nauk SSSR, Ser. Fiz. 41, 804 (1977)
6.25 Y. Ohmori, M. Jamaguchi, K. Hoshino, Y. Inushi: Jpn. J. Appl. Phys. 15, 2263 (1976)
6.26 K. Krätzig, H. Kurz: Ferroelectrics 13, 295 (1976)
6.27 W. Keune, S.K. Date, U. Gonser, H. Bunzel: Ferroelectrics 13, 443 (1976)
6.28 B. Dischler, A. Räuber: Solid State Commun. 17, 953 (1975)
6.29 H. Tsuya: Izv. Akad Nauk SSSR, Ser. Fiz. 41, 740 (1977)
6.30 W.J. Merz: Helv. Phys. Acta 31, 625 (1958)
6.31 S.G. Ellis, F. Herman, E.E. Loebner, W.J. Merz, C.W. Struck, G. White: Phys. Rev. 109, 1860 (1958)
6.32 P.S. Brody: Solid State Commun. 12, 673 (1973)
6.33 W. Jösch, R. Munser, W. Ruppel, P. Würfel: *Ferroelectrics: 4th International Meeting on Ferroelectricity* (Leningrad, 1977) In press
6.34 G. Chanussot, A.M. Glass: Phys. Lett. 59A, 405 (1976)
6.35 V.I. Belinicher, I.F. Kanayev, V.K. Malinovsky, B.I. Sturman: Avtometriia 4, 23 (1976)
6.36 V.I. Belinicher, V.K. Malinovsky, B.I. Sturman: Zh. Eksp. Teor. Fiz. 73, 692 (1977)
6.37 D. Johnston: J. Appl. Phys. 41, 3279 (1970)
6.38 G. Chanussot: Ferroelectrics 13, 313 (1976)
6.39 S.H. Wemple, M. DiDomenico, A. Jayaraman: Phys. Rev. 180, 547 (1969)
6.40 G.A. Alphonso, R.C. Alig, D.L. Staebler, W. Phillips: RCA Rev. 36, 213 (1975)
6.41 V.M. Fridkin, A.A. Grekov, P.V. Ionov, A.I. Rodin, E.A. Savchenko, K.A. Verkhovskaya: Ferroelectrics 8, 433 (1974)
7.1 A. Ashkin, C.D. Boyd, T.M. Dziedzic, R.G. Smith, A.A. Ballman, I.I. Levinstein, K. Nassau: Appl. Phys. Lett. 9, 72 (1966)
7.2 A. Ashkin, B. Tell, I.M. Dziedzic: J. Quantum Electronics, 3, 400 (1967)
7.3 F.S. Chen: J. Appl. Phys. 8, 3418 (1967)
7.4 I.I. Amodei: RCA Rev. 32, 185 (1971)
7.5 M.G. Clark, F.S. Di Salvo, A.M. Glass, G.E. Peterson: J. Chem. Phys. 59, 6209 (1973)
7.6 D. von der Linde, A.M. Glass, K.R. Rodgers: Appl. Phys. Lett. 25, 155 (1974)
7.7 R.L. Townsend, L.T. La Macchia: J. Appl. Phys. 41, 5188 (1970)
7.8 I.I. Amodei, D.L. Staebler, W. Stephens: Appl. Phys. Lett. 18, 507 (1971)
7.9 I.I. Amodei, D.L. Staebler, W. Phillips: Appl. Optics 11, 390 (1972)
7.10 D.L. Staebler, I.I. Amodei: Ferroelectrics 3, 107 (1972)
7.11 N.B. Angerb, V.A. Pashkov, I.M. Solovieva: Zh. Eksp. Teor. Fiz. 62, 1666 (1972)
7.12 H. Tsuza, Y. Fujino: Jpn. J. Appl. Phys. 12, 1896 (1973)
7.13 I.B. Thaxter, M. Kestigian: Appl. Optics 13, 913 (1974)
7.14 F. Micheron, C. Mayeux, I.C. Trotier: Appl. Optics 13, 784 (1974)
7.15 V.M. Fridkin, B.N. Popov, K.A. Verkhovskaya: Appl. Physics (1978) in press
7.16 J.P. Huignard, F. Micheron: Appl. Phys. Lett. 29, 591 (1976)

7.17 J.J. Amodei: Appl. Phys. Lett. 18, 22 (1971)
7.18 A.V. Guinsberg, K.D. Kochev, Yu.S. Kusminov, T.R. Volk: Phys. Stat. Sol. (a) 29, 309 (1975)
7.19 P.V. Lenso, E.C. Spencer, A.A. Bollmann: Appl. Phys. Lett. 11, 33 (1967)
7.20 A.M. Glass, D.H. Auston: Phys. Rev. Lett. 28, 897 (1972)
7.21 D. von der Linde, D.H. Auston, A.M. Glass, K.F. Rodgers: Solid State Commun. 14, 137 (1974)
7.22 I.F. Kanayev, V.K. Malinovsky: Fiz. Tverd. Tela (Leningrad) 16, 3694 (1974)
7.23 D. von der Linde, A.M. Glass: Appl. Phys. 8, 85 (1975)
7.24 H. Kogelnik: Bell Syst. Tech. J. 48, 2909 (1969)
7.25 M. DiDomenico, S.H. Wemple: J. Appl. Phys. 40, 720 (1969)
7.26 V.M. Fridkin, K.D. Kochev, Yu.S. Kusminov, K.A. Verkhovskaya, T.R. Volk: Phys. Stat. Sol. (a) 33, K137 (1976)
7.27 D.W. Vahey: J. Appl. Phys. 46, 3510 (1976)
7.28 R.L. Byer, C.B. Bounty: Ferroelectrics 3, 33 (1972)
7.29 D. von der Linde, A.M. Glass, K.F. Rodgers: Appl. Phys. Lett. 26, 22 (1975)
7.30 S.L. Hou, D.S. Oliver: Appl. Phys. Lett. 18, 325 (1971)
7.31 D. von der Linde, A.M. Glass, K.F. Rodgers: Appl. Phys. Lett. 25, 155 (1974)
7.32 D. von der Linde, O.F. Schirmer, H. Kurz: In *Ferroelectronics: 4th International Meeting on Ferroelectricity* (Leningrad, 1977) In press
8.1 V.M. Fridkin, I.I. Groshik: Appl. Phys. Lett. 10, 354 (1967)
8.2 A.A. Grekov, V.A. Lyakhovitskaya, A.I. Rodin, V.M. Fridkin: Dokl. Akad. Nauk SSSR 169, 810 (1966)
8.3 T. Mori, H. Tamura, E. Sawaguchi: J. Phys. Soc. Jpn. 20, 1294 (1965)
8.4 S. Kawada, M. Ida: J. Phys. Soc. Jpn. 20, 1287 (1965)
8.5 D.S. Liebeyman, M.S. Wechsler, T.H. Read: J. Appl. Phys. 26, 473 (1955)
8.6 T. Mori, E. Sawaguchi: Jpn. J. Appl. Phys. 6, 792 (1967)
8.7 T. Mori, E. Sawaguchi: J. Phys. Soc. Jpn. 25, 1195 (1968)
8.8 A.G. Chynoweth: Phys. Rev. 102, 705 (1956)
8.9 D. Meyerhoffer: Phys. Rev. 112, 413 (1958)
8.10 V.V. Ivanov, A.A. Bogomolov, V.M. Rudyak: Fiz. Tverd. Tela (Leningrad) 11, 3394 (1969)
8.11 A.A. Grekov, V.A. Lyakhovitskaya, A.I. Rodin, V.M. Fridkin: Fiz. Tverd. Tela (Leningrad) 10, 2239 (1968)
8.12 A.Yu. Kudzin, A.N. Sukhinsky, V.A. Butsichenko: Fiz. Tverd. Tela (Leningrad) 12, 2180 (1970)
8.13 C.E. Land: Ferroelectrics 7, 45 (1974)
8.14 W.D. Smith, C.E. Land: Appl. Phys. Lett. 20, 169 (1972)
8.15 C.E. Land, P.S. Peercy: In *Ferroelectrics: 4th International Meeting on Ferroelectricity* (Leningrad, 1977) In press
8.16 M.P. Michailov: C.R. Acad. Bulg. 21, 1029 (1968)
8.17 B.P. Grigas: Fiz. Tverd. Tela (Leningrad) 9, 2430 (1967)
8.18 V.P. Bender, V.M. Fridkin: Fiz. Tverd. Tela (Leningrad) 13, 614 (1971)
8.19 P.W. Forsberg: Phys. Rev. 76, 1187 (1949)
8.20 K. Imai, S. Kawada, M. Ida: J. Phys. Soc. Jpn. 21, 1855 (1966)
8.21 N.M. Bezdetny, G.Z. Gorbatov, A.Kh. Zeinally, N.N. Lebedeva, M.K. Sheinkman: Fiz. Tverd. Tela (Leningrad) 14, 924 (1972)
8.22 N.M. Bezdetny, A.Kh. Zeinally, N.N. Lebedeva, M.K. Sheinkman: Fiz. Tekh. Poluprovodn. [Sov. Phys. - Semicond.] 5, 1061 (1970)

Subject Index

MIX
Papier aus verantwortungsvollen Quellen
Paper from responsible sources
FSC® C105338

If you have any concerns about our products,
you can contact us on
ProductSafety@springernature.com

In case Publisher is established outside the EU,
the EU authorized representative is:
Springer Nature Customer Service Center GmbH
Europaplatz 3, 69115 Heidelberg, Germany

Printed by Libri Plureos GmbH
in Hamburg, Germany